PLANT GENETIC TRANSFORMATION TECHNOLOGY

PLANT GENETIC TRANSFORMATION TECHNOLOGY

Wagdy A. Sawahel
B.Sc., Ph.D., CBiol., MIBiol., EurBiol.
Genetic Enginneering and Biotechnology Division
National Research Centre
Cairo, Egypt

2026
Daya Publishing House®
A Division of
Astral International Pvt. Ltd.
New Delhi - 110 002

First Published, 1997
Reprinted 2026

ISBN: 978-81-7035-175-7 (Hardbound)

Published by : **Daya Publishing House®**
A Division of
Astral International Pvt. Ltd.
– ISO 9001:2008 Certified Company –
4760-61/23, Ansari Road, Darya Ganj
New Delhi-110 002
Ph. 011-43549197, 23278134
E-mail: info@astralint.com
Website: www.astralint.com

Printed at : **Replika Press Pvt. Ltd**

ACKNOWLEDGEMENTS

I wish to express my appreciation to many colleagues from the various institutions, departments and countries all over the world who have provided me with valuable first-hand information, research finding and various forms of assistance. Notable among these are the following :

Drs K. Fukui (Japan), J. Paszkowski (Switzerland), A. Cuming (England), G. Sastry (England), K. Richard (France), R. Sven (Sweden), G. Weber (Germany), A Zentgraf (Germany), C. Amla (India), L. Erickson (Canada), C. Jacop (Belgium), Y. Yamada (Japan), and C. Morikawa (Japan).

PREFACE

In recent years, and throughout the world, the ability of scientists to analyse gene structure and regulation has become more and more advanced. Consequently, an extremely high level of skill has now been acquired, and a great quantity of new knowledge has arisen within this area of research. This is, without doubt, largely due to the development of gene transfer to plants as an important tool for use in studies in basic and applied plant biology.

The abundance of new knowledge is in part due to the progress that has been made in the development of new systems for introducing foreign DNA into plant cells and the advances that have been made in the studies of transient and integrative transformation in both homologous and heterologous systems. One can therefore expect that, in the future, gene transfer, in combination with strategies for gene isolation, DNA modification, *in situ* hybridization *etc.*, will play an essential role in the analysis and regulation of plant development.

As the reader may be aware, a large number of well established, potential, plant transformation systems exist which have become routine for certain applications. As well as these well established systems there are a wide variety of systems which still remain speculative because their value has not yet been completely proven. Other systems, although well documented, have been constrained by limited practical utility. In addition to all of these factors, another point is that although a large number of gene transfer systems have yielded interesting results, they have often been published in such a way that it has been extremely difficult to assess the reliability and efficiency of the system, and to judge whether or not integrative transformation has occurred. In other words, although an abundance of knowledge exists within the area of gene transfer, its documentation remains disjointed.

In writing this book, I have tried to draw together all the available information on gene transfer systems in plants, including, wherever possible, all the most up-to-date systems and research. In doing so I have aimed to achieve two goals. First of all, I have attempted to review the whole of this exciting area of research in order to describe what has been achieved so far, where the limitations lie and what might still be achieved in the future. Secondly, I have tried to produce a basic reference book for use by specialists and non-specialists alike.

Finally, I sincerely hope that this book provides the reader with a comprehensive review of all the work that has been published to date within the area of gene transfer technology. I have tried, to the best of my ability, to give research credit to those researchers and scientists who deserve recognition for the work that they have done in this field.

Wagdy A. Sawahel

Contents

INTRODUCTION

The genetic revolution as such, refers to gene splicing, gene sequencing, gene synthesis and gene transfer. This book will concentrate on the latter topic, namely gene transfer.

For the purpose of this book "gene" will be generally defined as any of several small parts of the material at the NUCLEUS (= centre) of a cell, that control the development of all the quantities in a living organism which have been passed on from its parents. More specifically, it will be defined as a segment of DNA that codes for an RNA and/or polypeptide molecule. In general, transfer refers to an act, or process, of transferring and in this book it specifically refers to the introduction of foreign, genetic material into any recipient cell, tissue or organelle.

As a result of the genetic revolution and the interest which it generated within this field, gene transfer systems have been developed for many important species, and, routine transformation of certain bacterial, fungal, plant and animal cells now take place in laboratories all over the world. However, although a complex variety of transformation systems exist, many of them are ineffective. Consequently, even where there are reports of gene transfer into a given species, or strain, scientists occasionally find that many of the systems used are impractical. A number of reasons account for this. A system may, for example, be unreproducible, may not have the necessary delivery efficiency, or, may not be rapid or cheap enough. In contrast to this, highly effective methods also exist but they too have their problems. For example, they may require infectious agents which in itself may present a number of associated biological and legal complications.

For many years, scientists have been interested in using gene transfer and manipulation techniques for the predictable modification of plant genomes. They have particularly concentrated on plants which have had a certain value, or importance, either for economic or nutritional reasons. In addition to searching for gene transfer system, scientists have aimed at achieving an ideal, gene transfer system which can be generally applied.

Despite the existence of a variety of ideal transformation systems, no universal, ideal gene transfer system, *i.e.* one that can be generally applied to all plant species has, as yet, been found. This still remains a "stumbling block" for scientists throughout the world but hopefully this will be solved within the near future.

CHAPTER I

GENETIC TRANSFORMATION

1. History & Definition

The phenomenon of transformation holds a special place in genetics as the process in which the molecular basis of heredity was first shown to be DNA. Transformation was discovered by Griffith (1928) in the course of experimenting with *Diplococcus* (or *Streptococcus*) *pneumoniae* (the pneumococcus). He found that the smooth capsule of polysaccharide surrounding the cell wall of a virulent strain could be inherited by an avirulent non-encapsulated strain with which it was mixed, even though the cells of the virulent strain had first been killed by heat. Subsequently, it was shown that cell-free extracts of the virulent strain could bring about transformation of the other strain, and ultimately Avery *et al.* (1944) purified the substance responsible and showed that it was DNA.

Genetic transformation is a process by which a cell takes up naked DNA from the surrounding medium and incorporates it to acquire an altered genotype that is heritable (Wilson, 1985). Traditionally, the term transfection refers to the infection of a host cell with naked viral DNA by means of the cell's transformation machinery. Plasmid DNAs are similar in many respects to viral DNAs and seem to be handled by cells in the same fashion during transformation. Therefore, some scientists use the term transfection to refer to transformation with plasmid or viral DNAs (Brisson *et al.*, 1984).

In recent years the term transformation has been extended to include uptake of any DNA molecule by any type of cell, regardless of whether the uptake results in a detectable change in the cell or whether the cell involved is bacterial, fungal, animal or plant (Walden, 1988).

2. Gene Transfer Systems

For many years, scientists have been interested in using gene transfer and manipulation techniques for the predictable modification of plant genomes. They have particularly concentrated on plants which have had a certain value, or importance, either for economic or nutritional reasons.

An ideal gene transfer system which can be generally applied would satisfy a number of requirements. First of all, it should allow for transformation of the already transformable cells, tissues and species, but it would also be faster, simpler and more direct than existing methods. Secondly, it should facilitate gene transfer to cells, tissue and species for which no other transformation method has, as yet, been defined. Thirdly, it should be broadly applicable and would enable researchers to use a single apparatus (and the same basic protocol) for all types of gene transfers (Potrykus, 1990).

2.1. Natural Transformation Systems (vector system)

(A) *Agrobacterium tumefaciens* and *Agrobacterium rhizogenes*

The pathogenic bacteria, *A. tumefaciens* and *A. rhizogenes*, are capable of delivering DNA from their Ti or Ri plasmids respectively into cells of a wide variety of dicotyledonous and some monocotyledonous plants. Although both systems rely on a similar transformation mechanism (Hoekema *et al.*, 1986 & Slightom *et al.*, 1985) the effects of the Ri and Ti T-DNA are different.

Agrobacterium tumefaciens–mediated gene transfer is a mature and well established method and was at one time the only way to introduce foreign genes into plants. This technique has many advantages as it does not require protoplasts and transformation frequencies and integration are probably directly comparable. There may also be some advantage connected with integration via the T-DNA border (s). Until now there is little information concerning stability or instability of genes in consecutive sexual generations. One of the disadvantages of *Agrobacterium* is the host range which excludes, for example, tne graminaceous monocots. In

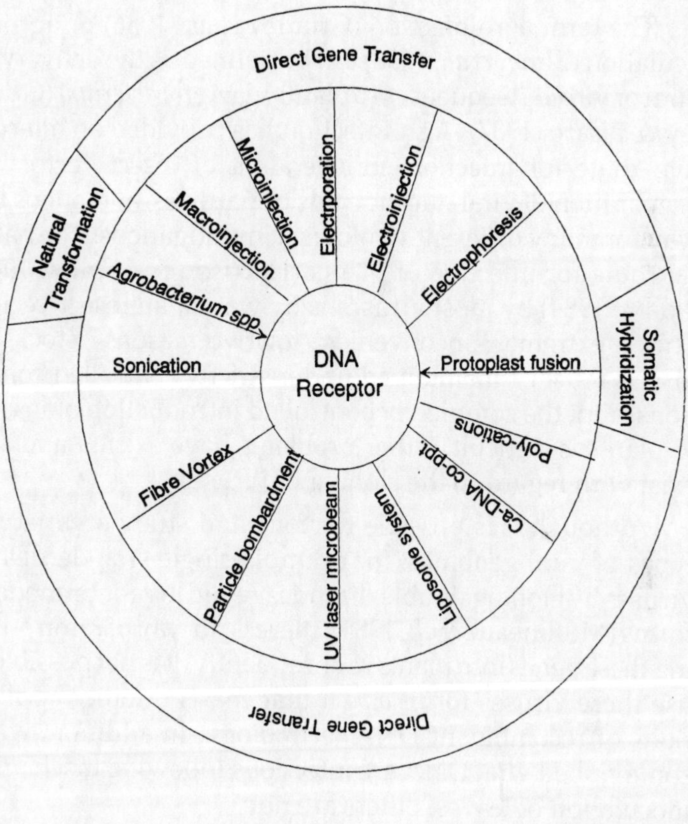

Fig. 1: Gene Transfer Systems

addition, it is not yet known how T-DNA crosses the plant cell wall and reaches the nucleus and integrates into the host genome.

(B) Agroinfection

The term agroinfection (Grimley *et al.*, 1986) or agroinoculation (Elmer *et al.*, 1988) can be defined as the delivery of viral or viroidal sequences to plants via *Agrobacterium* (Grimley & Bisaro, 1987). Agroinfection has provided an alternative route for infection, independent of insect vectors or preparation of viral nucleic acids or particles, and opens the way for many different virological investigations. The natural route for infection of plants (Harrison & Murant, 1984; Adam, 1984) by most viruses is by way of an insect vector. This is extremely inconvenient for two reasons : stocks of insects must be maintained under strictly controlled conditions, and, the completely controlled introduction of insects to plants is difficult. Other problems have existed and still exist with regard to the cloning of viruses.

Although it is possible to clone and study a variety of types of viral genomes, for example single-stranded DNA viruses by cloning double-strandes replicative-intermediate forms (Mullineauz *et al.*, 1984), there is no way of reintroducing these genes into plants. Consequently, it is not possible to use these viruses for *in vivo* mutagenesis or other applications of recombinant DNA technology. In addition, while some cloned viral DNA are infectious following mechanical inoculation of leaves, others are not.

Agroinfection can be used to approach several important and basic questions in plant molecular biology. For example, the plant *Agrobacterium* interaction can be analysed using the virus as a marker for T-DNA transfer or, viral biology can be investigated using *Agrobacterium* as a delivery vehicle for the virus (Rogers *et al.*, 1986 ; Mullineaux *et al.*, 1988).

Plants, transgenic for viral constructs, offer possibilities for studying recombination (Grimsley *et al.*, 1986b), plant protection mechanisms (Gerlach *et al.*, 1987 ; Harrison *et al.*, 1987) and the development of high copy number plant vectors (Brisson *et al.*, 1984 ; Hayes *et al.*, 1988b).

2.2. Direct Gene Transfer (vector-free systems)

The host range of *Agrobacterium tumefaciens* is limited. While more than twenty crop species have now been transformed using this system, it has not proved to be immediately useful for a number of dicots (although the number of successes in this respect is increasing rapidly), and has been found to be completely ineffective for transformation of the commercially important cereal crops such as maize, wheat and barley, though limited success has recently been achieved for rice (Raineri *et al.*, 1990). In response to the observed limitations in the application of *Agrobacterium*, efforts have been directed to the development of alternative, vector-free or direct transfer systems for plant transformation.

Direct gene transfer describes the phenomenon of uptake of genes into plant cells through the naked plasmid membrane or the cell wall independent of a biological vector and their functional integration into the host genome (Potrykus *et al.*, 1985). The advantages of direct gene transfer are that the form of the DNA which is applied to protoplasts is controlled entirely by the person carrying out the experiment and not by an intermediate biological vector. Therefore, methods such as co-transformation (Fromm *et al.*, 1986), transformation with clones or whole genomic DNA and various manipulations of DNA before application to the plant cells, can easily be carried out (Fromm *et al.*, 1985).

3. Genetic Transformation Strategy

3.1. Biological Parameters

The biological parameters play a role in gene transfer experiments. Some of these parameters are, according to Potrykus *et al.* (1990), as follows :

(1) Plant cells differ in their capacity to respond to external triggers, a phenomenon termed competence. Plant cells are composed of cells competent for different responses. A very small minority of cells in plant tissue are competent for transformation (either transient or integrative), others are competent for either regeneration or for transformation and a variable part of the cells are non-competent or terminally differentiated. The relative composition of the cell populations in tissues is determined by the genotype, the type of organ, the position of the organ, its developmental state, and even the individual history of the experimental plant or organ. The production of transgenic plants requires an efficient gene transfer system as well as cells competent for integrative transformation and regeneration.

(2) Plant cell walls are efficient barriers and traps for DNA molecules of the size of a functional gene. However, genes can be transferred into cells across walls by using *Agrobacterium* or biolistic transformation, or, if the cell wall is removed, through the plasmalemma.

(3) Gene transfer *via* DNA virus, as described for the first time for Cauliflower Mosaic Virus (Brisson *et al.*, 1984) has a specific merit based on the systematic spread of the virus throughout an entire plant and the multicopy situation per cell infected. However, it has limitations based on the host range, it does not integrate into the host genome and lack of transfer to sexual offspring.

(4) Non-viral DNA is restricted to the cell to which it was delivered and can integrate into the host genome.

3.2. Requirements for Genetic Transformation

In order for a gene transfer event to be successful, various requirements must be met. First, the foreign DNA must enter the cell, passing through the physical barriers posed by the cell wall and/or cell membranes. This is accomplished in different ways for each gene transfer system. Once the DNA has penetrated the cell, it must remain fairly intact. This entails avoidance of degradation by nucleases. Also, the DNA must arrive at a region where it can function. Having reached a suitable location, the DNA must be transcribed and translated. In some cases, the incoming DNA may actually integrate into the host chromosome where it is presumably transcribed in the same manner as the regular chromosomal genes. In other cases, the DNA remains as unintegrated and it is transcribed independently.

The introduction of foreign genetic material into recipient cells requires the following (Fig. 2).

(a) Source of the hybrid-selectable gene (DNA Donor). Several hybrid marker genes for use in plant cell transformation, using *Agrobacterium tumefaciens*-mediated transformation, have been described in the last ten years (Herrera-Estrella *et al.*, 1983). The elements necessary in such a construction can be summarized as follows :

— Plant gene expression signal, *i.e.*, promoter and terminator regions for an RNA, which are best derived from a constitutively and highly expressed plant or plant viral gene.

— A protein-coding region joined precisely to the above expression signals which, when expressed, will give an active product which allows easy selection at the plant cell level, *e.g.*, detoxification of antibiotics lethal for plant cells.

— For DNA-mediated transformation, a region on the bacterial vector plasmid which allows recombination into the plant genome without disruption of the selectable gene.

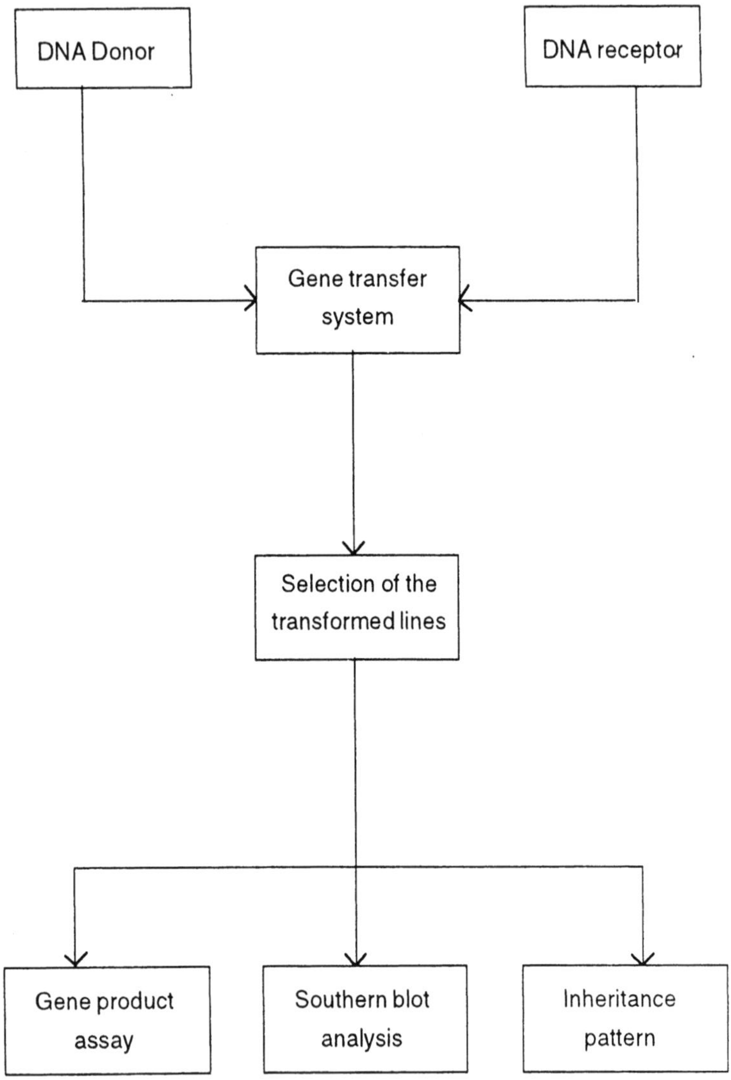

Fig. 2: Genetic transformation strategy

(b) DNA receptor.

There are a number of DNA receptors which can be used such as protoplasts, cells, tissues, seeds, embryos, ovaries, pollen, *etc.* However, the particular choice depends upon the gene delivery system to be used as well as the regeneration capacity within the plant to be transformed.

(c) Gene transfer system.

See Section 2.

(d) Confirmation of the transformation event.

The criteria for confirmation of a transformation event should be the following :

(A) Phenotypic change to resistance in a selection scheme which is proven to be "clean". Ideally the level of selection should permit recovery of only transformed clones. Therefore, resistant clones should only appear after transformation with the correct vector and not in any control treatments (Fig. 3).

(B) Expression of the foreign DNA at the RNA/protein level. Activity assay for the product of the transforming gene will of course depend on the expected product (Fig. 4).

The use of fusions between a gene of interest and a reporter gene that encodes an easy detectable product (not found in the organism being studied) facilitate studies of the mechanism of transformation. So far, six reporter genes have been used in studies of gene expression in higher plants. These include the *Escherichia coli* β-galactosidase (Helmer *et al.*, 1984) ; the *Agrobacterium tumefaciens* Ti-plasmid-encoded genes nopaline synthase (Herrera-Estrella *et al.*, 1983 & Bevan *et al.*, 1983) and octopine synthase (Jones *et al.*, 1985); the chloramphenicol acetyl transferase (*CAT*) (Gorman *et al.*, 1982) ; the neomycin phosphotransferase (*NPT II*) (Resis *et al.*, 1984) and the firefly luciferase (Ow *et al.*, 1986).

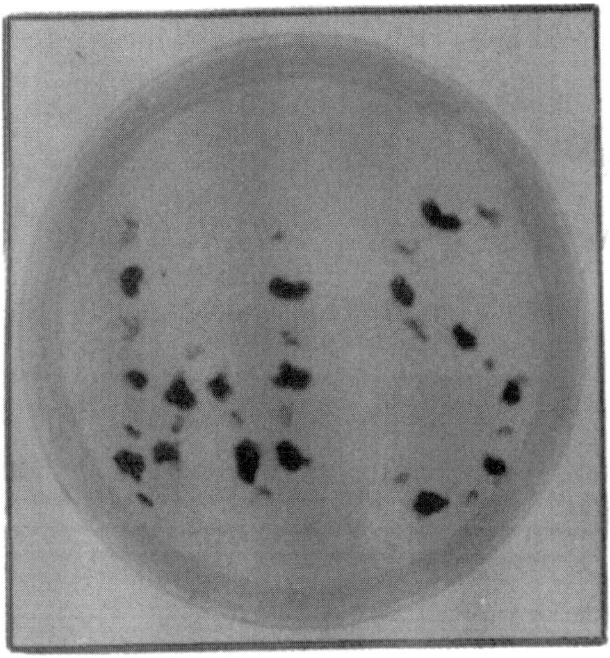

Fig. 3: Stable transformants (green clones) showing unrestricted growth on a selective medium after growing for four weeks in the absence of selection : the untransformed clones (brown clones) had died.

Fig. 4: Gene product assay

In recent years, an easy and sensitive enzymological reaction using the *Escherichia coli* β-glucuronidase, encoded by the *uidA* gene, as reporter enzyme for the detection of transformed tissues (Jefferson *et al.*, 1986) has extensively been exploited in plant genetic engineering. The coding region of this gene, also called GUS gene, is a useful marker because it is able to cleave various commercial substrates for a variety of spectrophotometric, fluorimetric and histochemical assays (Fig. 5). Incubation with an indigogenic substrate and preparation for light microscopy enable the visualization of transformed cells (Taylor & Vasil, 1991) and allows investigation of gene expression, tissue-specific regulation of gene constructs (Terada & Shimamoto, 1990) and protein targeting (Daniell *et al.*, 1991). Furthermore, as β-glucuronidase is generally not present in plants (Jefferson, 1987), with only a few exceptions in some tissues (Hu *et al.*, 1990), the GUS gene offers an advantage compared to another reporter gene, *LacZ*, encoding β-galactosidase, which can be used in the same way, and which requires pretreatments of tissues in order to eliminate endogenous β-galactosidase (Teeri *et al.*, 1989).

(C) Molecular analysis of the DNA of transformed clones and regenerated plants. DNA of transformed cell lines, regenerated plants, and their progeny should be analyzed using standard Southern blot techniques. It should be possible to prove unequivocally the presence of the transforming DNA in high molecular weight nuclear DNA of transformed lines in an integrated form in the absence of any hybridization to DNA from control untransformed lines.

4. Arrangement of foreign DNA in the plant genome

Potrykus *et al.* (1985) have reported that detailed genetic and molecular analysis of transformed plants and their progeny revealed the following features : Integration of plasmid molecules into the nuclear genome occurs in a

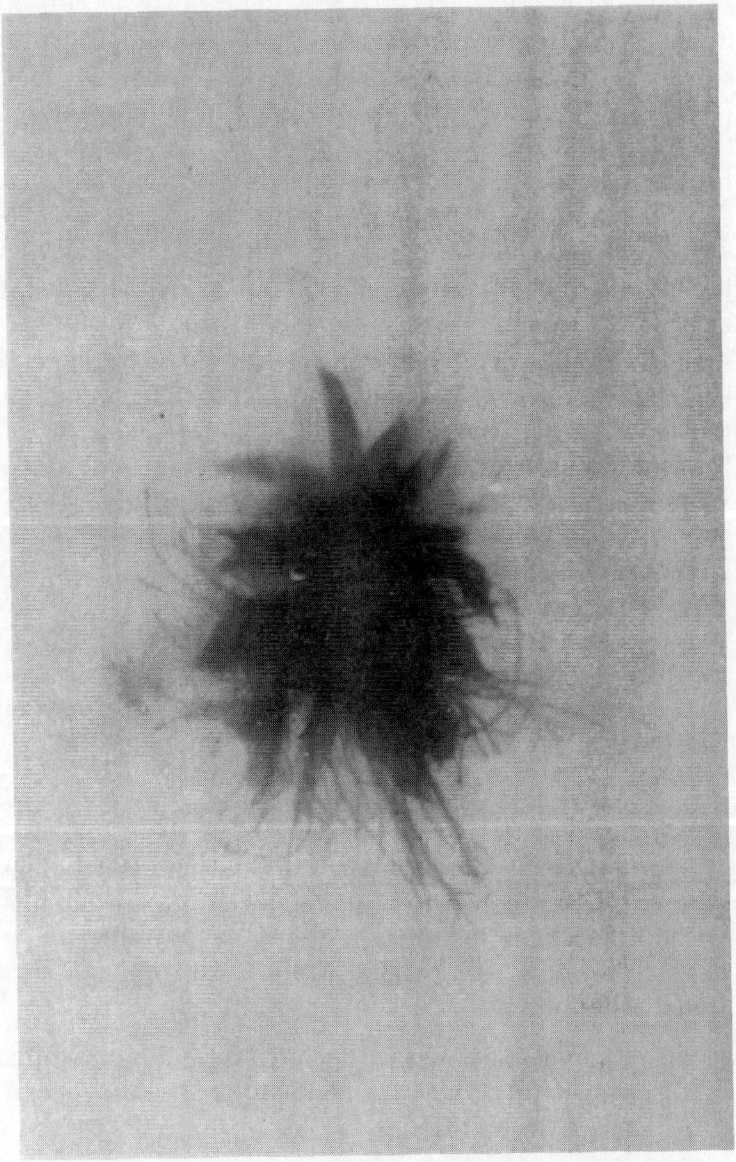

Fig. 5: GUS expression pattern in stable transformant

fashion similar to that described for DNA transfection of animal cells. Paszkowski *et al.* (1984) ; Potrykus *et al.* (1985a, b) and Hain *et al.* (1985) have found integration of several copies of the transforming plasmid DNA. Potrykus *et al.* (1985) have reported that analysis of several independent transformed cell clones and plants regenerated from them resulted in an unexpected finding : the integrated DNA copies were not randomly distributed on several chromosomes but were integrated in only one physical location in the tobacco genome. Although integration always occurs at a single location, this locus differs in different transformed lines (Potrykus *et al.*, 1985a). The pattern of integration and the degree of rearrangement is dependent upon the species, the developmental state of the recipient protoplasts and the transformation method applied (Paszkowski *et al.*, 1984 ; Potrykus *et al.*, 1985 b, c). However, Christou *et al.* (1988) have reported that the general feature of DNA integration and gene copy number were found to be similar in tissue derived from both the PEG method and particle bombardment. This suggests that although the delivery of DNA by these two methods is different, the same pathway of recombination and integration operate on the DNA once it has been delivered into the cell.

4.1. Stability of the foreign gene

Potrykus *et al.* (1985a) have studied the fate of the foreign gene through three sexual generations and for 18 months without selective pressure. They have demonstrated a high degree of mitotic and meiotic stability. This may be illustrated as follows :

(a) Of sixty subclones regenerated from 2 transformed clones, fifty-seven maintained the gene under non-selective conditions.

(b) Regenerated plants are neither chimeras nor mosaics.

(c) The gene is regularly transmitted to sexual offspring.

(d) The inheritance of the gene follows strictly Mendelian rules.

(e) The gene has been maintained over three sexual generations.

(f) The gene is not only stably maintained when present in homozygous situation, but also when in a heterozygous state (30 R2 seedling populations produced by self-crosses of 30 flowers from primary, secondary and tertiary inflorescences of a heterozygote transformant, all segregated 3 : 1).

The genetic data are supported by the inheritance of a transformant-specific hybridization pattern in Southern blot analysis (Potrykus *et al.*, 1985b).

4.2. Modes of genetic recombination

One of the most useful parameters for distinguishing different modes of recombination at this stage remains the base sequence homology, or lack of it, between the two DNAs which react to form a recombinant molecule or molecules (Wilson, 1985). This is due in part to the lingering obscurity of the sequence and number of the enzymatic functions involved in genetic recombination (Clark, 1973).

Recombination events that have been characterized can be grouped conveniently into three classes :

A–General recombination

When the extent and degree of homology between parental molecules are great, recombination in most organisms is a fairly frequent event, occurring largely at random regions where the two homologous parental DNAs are (somehow) brought into register.

This class has been referred to by numerous synonyms, such as recombination, generalized (Dove, 1968), normal

(Franklin, 1971), equal (Campbell, 1964), genetic (Hotchkiss, 1974), rec-dependent (Franklin, 1971), chromosomal (Gottesman & Yarmolinsky, 1968), vegetative [when referring to phage recombination] (Hotchkiss, 1974), legitimate (Falkow, 1975), non-specific (Gottesman & Yarmolinsky, 1968), equational (Campbell, 1960) or mating (Gottesman & Yarmolinsky, 1968).

B–Site-specific recombination

Recombination events occur between DNA molecules between which is very little, if any, homology. When the location of recombination event is a highly preferred crossing over location on one (single-site specific) or both (double-site specific) of the parental DNA molecules.

The process of site-specific recombination has been referred to in numerous ways, such as, specific (Ginger & Echols, 1967), special (Singh & Beckwith, 1966), localized (Gottesman & Yarmolinsky, 1960), additive (Schwesinger, 1977) and non-homologous (Kornberg, 1974).

C–Illegitimate recombination

Recombination events take place at a low frequency at apparently random sites between regions of DNA that exhibit very little, if any, homology.

This class has been referred to by numerous synonyms, such as improper (Stent, 1971), Unusual (Franklin, 1971), non-homologous (Kornberg, 1974), non-equational (Campbell, 1960) and unequal (Campbell, 1964).

Although integration of transforming DNA into the homologous chromosomal DNA occurs efficiently in yeast, other fungi and *Diclyostelium discoideum* (Hinnen *et al.*, 1978; De Lozanne *et al.*, 1987 ; Miller *et al.*, 1987). In plants whatever the method, incoming DNA integrates at random locations (Potrykus *et al.*, 1985b ; Ambros *et al.*, 1986 ; Nagy *et al.*, 1986 ; Wallroth *et al.*, 1986). The high transformation frequencies reported for DNA molecules with no homology to plant genome (Shillito *et al.*, 1985; Negrutiu *et al.*, 1987) indicate efficient non-homologous "illegitimate" integration of the foreign DNA into the plant genome.

Foreign DNA introduced into the nuclei of higher eukaryotes generally integrates into non-homologous sequences, even when the introduced DNA has homology to endogenous sequences (Lee *et al.*, 1990). This hampers attempts to direct introduced DNA sequences to homologous locations in the genome (Lee *et al.*, 1990). Direct integration, *via* homology is recombination, of introduced DNA sequences at sites in the gene where homologous sequences exist (gene targeting) would be a powerful technique for the molecular genetic study of higher organisms.

Gene targeting *via* homologous recombination was recently reported in tobacco cells (Paszkowski *et al.*, 1988). The target was a stably integrated defective NPT II gene that was present in multiple copies at one locus. The exogenous DNA in these experiments was introduced by polyethylene glycol-mediated transformation, and the ratio of homologous events to random insertion events was 10^{-4}.

5. Genetic transformation approaches

Davey *et al.* (1980), Krens *et al.* (1982), and Paszkowski *et al.* (1984) were the first to devise new methods of direct gene transfer, independently of the *Agrobacterium* technique, which opened a wide new approach for plant transformation systems. The number of genes or size of a DNA fragment that can be transferred either by using natural transformation system or direct gene transfer at one time is limited by the availability of unique restriction enzyme sites following each insertion. However, for certain purposes, it could be advantageous to transfer several genes into a plant genome. Two approaches for overcoming this limitation are cotransformation and retransformation.

Several genes can be introduced into a plant cell during sequential transformation steps (retransformation) if an alternative system to select transformants is available for the second round of transformation. Sequential transformation steps using kanamycin and hygromycin selection respectively, have produced transgenic tobacco plants using both sense and antisense construction of either the nopaline synthase gene (Sandler *et al.*, 1988) or the gene for chloramphenicol acetyltransferase (Delauney *et al.*, 1988).

Cotransformation with more than one T-DNA has been accomplished by incubating plant cells either simultaneously with two different strains of *Agrobacterium*, each with different Ti-plasmid (Depicker *et al.*, 1985 ; Petit *et al.*, 1986 ; Simpson *et al.*, 1986 ; McKnight *et al.*, 1987 ; Prosen & Simpson, 1987), or with one strain of *Agrobacterium* containing two T-DNAs on separate plasmids (de Framond *et al.*, 1986 ; Hamill *et al.*, 1987).

Potrykus *et al.* (1985a) proposed in a review a large list of experimental opportunities for direct gene transfer. One of them is the cotransfer of an easily selectable marker gene with non-selectable gene. The fact that efficient direct gene transfer to plant protoplasts is possible by the addition of DNA, without a bacterial intermediate, suggests that they take up and integrate DNA in a non-selective way. This should led to uptake and integration of mixtures of different DNA molecules added to the protoplasts at the same time (Paszkowski *et al.*, 1984). It has shown that cotransformation can be obtained either by mixing the two plasmids containing the selectable marker and the non-selectable gene (Schocher *et al.*, 1986), or by constructing a vector harbouring the two genes (Uchimiya *et al.*, 1986). Pietrzak *et al.* (1986) used two bacterial antibiotic resistance genes on two separate plasmids to transform *Nicotiana tabacum* protoplasts.

Czernilofsky *et al.* (1986a, b) have shown that two physically unlinked but partly homologous plasmids were physically linked after integration into the plant host genome. It has been shown that the direct genetic transformation of protoplasts with two unlinked genes results in transformants that to a high percentage have both genes integrated (Christou & Swain, 1990). Co-integration and co-segregation of the plasmid and carrier DNA sequences has also been shown to occur (Peerbolte *et al.*, 1985 ; Jongsma *et al.*, 1987). In cotransformation experiments with two unlinked plasmid, 20-80 per cent of the clones that express one of the genes also expresses the other gene (Schocher *et al.*, 1986). This high efficiency of cotransformation provides a simple method for introduction of non-selectable gene into plant genomes (Shimamoto *et al.*, 1989).

6. Classes of transformants

The introduction of DNA molecules into eukaryotic cells leads to transient, unstable as well as stable gene expression (Fig. 6).

A–Stable transformants

Stable gene expression results from the expression of transferred DNA molecules which have been inserted, by recombination events, into the recipient cell chromosomal DNA. Genes located in such transferred DNA molecules follow the same pattern of behaviour as any other chromosomal gene (Ballas *et al.*, 1987), *i.e.*, there is no chance of losing the foreign DNA through asymmetric segregation during mitosis. This means, the presence of an appropriate stable gene product whether the stable transformants are under selection or non-selective-conditions.

Classes of transformants

Transient

"The expression of transferred DNA molecules within hours or days followed by subsequent decline. Such expression is assumed to be due to the presence of the transferred DNA, that is biologically active only for a limited time, as an unintegrated molecule within the cell nucleus."

Integrative (stable)

"Genetic modification of an organism mediated by the integration of foreign DNA sequences into the host genome."

Characters :–

- Mitotic stability
- Mendelian segregation of the new character
- Molecular evidence for the presence of hybrid fragments between host and foreign DNA

Replicative (unstable)

"Genetic transformation of an organism associated with extrachromosomal maintenance of efficiently replicated DNA sequences."

Sequence–dependent

DNA replication is initiated at specific bacterial sequences.

Sequence-independent "Autonomous replication"

DNA replication is initiated at non-specific sequences within the transforming plasmid.

Characters :–

- High transformation frequencies.
- Often high copy number of the transforming sequences.
- Mitotic instability of the character associated with mosaicism in the trnasformant. *i.e.* under non-selective conditions the plasmid suffers significant loss.
- Non-mendelian transmission of the new character.
- A tight correlation between phenotypical and physical loss of the genetic marker.
- After many generations of growth the plasmid DNA molecules may become integrated into the host chromosome.
- The existence of extrachromosomal primid DNA *i.e.* it should be possible to transfer the plasmid to *E. coli*.
- Under selective conditions the grcwth rate is slow *i.e.* generation times are increased.

Note :– In plants, neither sequence specific nor autonomous replicative transformation has yet been reported.

Fig. 6: Classes of transformants

Table 1 : Replicative transformation systems

Sequence-dependent	*Sequence-independent.*
Leishmania major	*Paramecium tetraurelia*
(Kapler *et al.*, 1990)	(Gilley *et al.*, 1988)
Tetrahymena thermophila	*Dictyostelium discoideum*
(Yu & Blackburn, 1990)	(Delozanne & Spudich, 1987)
Saccharomyces cerevisiae	*Caenorhabditis elegans*
(Newlon, 1988)	(Mello *et al.*, 1991)
Schizosaccharomyces pombe	*Neurospora crasse*
(Hayles & Nurse, 1992)	(Hughes *et al.*, 1983)
Phycomyces blakesleeanus	*Xenopus laevis*
(Revuelta & Jayaram, 1986)	(Marini *et al.*, 1988)
	Schizosaccharomyces pombe
	(Heyer *et al.*, 1986)
	Zebra fish
	(Stuart *et al.*, 1988)
	Sea urchin
	(McMahon *et al.*, 1985)

In some species, it has been found that multiple copies of the transfected plasmid had integrated at single locus. According to Perucho *et al.* (1980), two routes are possible which could lead to the integration of multiple copies of the plasmid into chromosome.

(a) A single copy of the plasmid could integrate at a locus in a chromosome and then serve as a site for the integration of the plasmid by homologous recombination.

(b) Concatemerisation of plasmid molecules could occur following transformation extrachromosomally by a process of intermolecular ligation "package of exogenous DNA" and then the concatemer integrates into the chromosome by a single recombinant event, *i.e.*, concatemerisation is a far more rapid event than chromosomal integration. This possibility requires that plasmid molecules are able to recombine with one another extra-chromosomally.

B–Transient transformants

Transient gene expression refers to the expression of transferred DNA molecules within hours or days following the transformation process followed by subsequent decline. Such expression is assumed to be due to the presence of the transferrred DNA, that is biologically active only for a limited time, as unintegrated molecules within the cell nucleus (Ballas *et al.*, 1987). The copy number of a gene within the cell nucleus should be the limiting factor for this type of transient expression (Topfer *et al.*, 1988).

C–Unstable transformant

Genetic transformation of an organism associated with extrachromosomal maintenance of efficiently replicated DNA sequences (Table 1). For example, transformation of Mammalian cells with metaphase chromosomes produced stable and unstable transformants (McBride & Ozer, 1973). Unstable transformants lose the transferred gene if selection pressure is removed. Unstable transformants fall into two different classes (Klobutcher *et al.*, 1980). Slow-loss transformants have a cytologically visible donor chromosome fragments with a centromere-like region and a low level of donor gene activity. Fast-loss transformants do not have a visible donor chromosome fragment but do have a high level of donor gene activity. The high level of gene activity in the latter class is probably due to an amplification of gene sequences (Scangos *et al.*, 1979). Both the high gene copy number (Scangos *et al.*, 1979) and the high level of gene activity (Degnen *et al.*, 1977) revert to normal when unstable transformants become stable, a rare event occurring in about $1/10^5$ cells/generation (Degnen *et al.*, 1976). Stability is achieved by apparently random integraion into a host chromosome (Klobutcher *et al.*, 1980; Athival & McBride, 1977; Klobutcher & Ruddle, 1979). Thus the transferred gene in unstable transformants is presumably propagated on an autonomously replicating element (Willecke & Ruddle, 1975), analogous to a bacterial plasmid, which tends to be better controlled by the cell with respect to copy number and segregation if it has a centromere-like region (Klobutcher *et al.*, 1980; Klobutcher & Ruddle, 1979).

7. Inter-transformant variability

Variability in gene expression has been observed between transformants for the majority of transformed plants (Jorgensen, 1990). Examination of independently transformed plants has shown a high degree of variability [up to 100–fold] in the expression level of the introduced gene (Jones *et al.*, 1985). In addition, some individual transgenic plants do not express the marker gene as expected (Budar *et al.*, 1986; Muller *et al.*, 1987). One exception to this phenomenon is the expression of a soybean conglycinin gene which varied only two fold in transgenic *petunia* plants (Nagy *et al.*, 1985).

It has been suggested that a number of factors may cause such inter-transformant variability, none of which fully explains the phenomenon (figure 7). These include the following :

7.1. Gene dosage (copy number)

Copy number of the introduced DNA varies among transformants. However, it has been shown that there is no positive correlation between increased copies at a given locus and increasing expression of the transgene (Shirsat *et al.*, 1989). In contrast, there are indications that single copies result in higher expression (Odell *et al.*, 1987).

7.2. Integration sites (position effect)

The insertion of the transgene is random within the plant genome and the activity of the introduced genes may be affected by adjacent DNA. However, this has been found in many instances not to be the simple, exclusive cause of variable expression (Dean *et al.*, 1988).

7.3. Truncation, rearrangement or repetition of the transgene

It has been shown that there has not always been a direct correlation between low gene expression and deletions, rearrangement (Deroles & Gardner, 1988), or inverted repeats (Jones *et al.*, 1987).

7.4. Epigenetic effects

It is now well established that methylation of DNA can exert its expression and can be inherited in a fairly stable manner (Holliday, 1987). Such epigenetic effects may contribute to variation between transformants. Primary regenerants with up to 16–fold differences in gene expression have been reported among genetically identical clones (Jones *et al.*, 1985).

7.5. Plant to plant physiological variability

Spatial and temporal differences in gene expression can occur if plant material is not physiologically equivalent (Williamson *et al.*, 1989). It has been suggested that the higher degree of variability in expression of the *rbcS* genes compared to storage protein genes is partially due to the greater difficulty of physiological standardization of leaves than developing cotyledons (Jones *et al.*, 1987).

7.6. The effect of the environment

Such an effect is especially important if the promoter

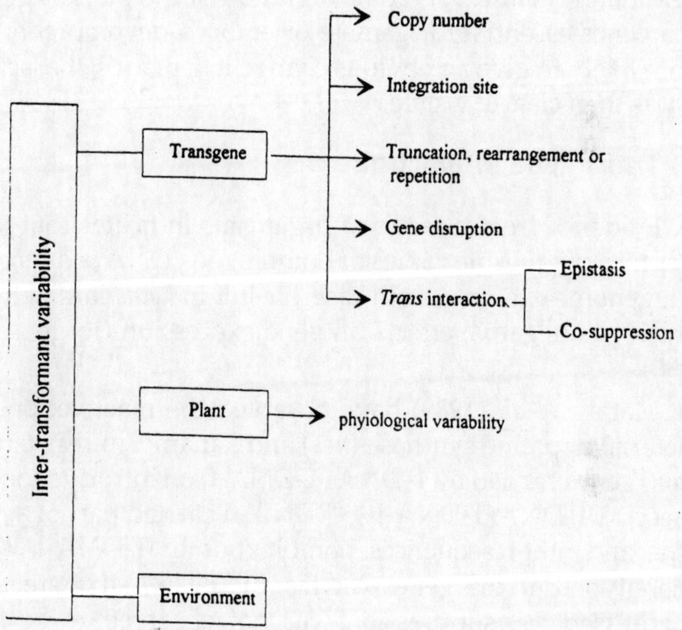

Fig. 7 : Possible causes for inter-transformant variability

used to control the expression of the transgene is affected by environmental stimuli such as light, heat, anaerobic stress and wounding (Benfey & Chua, 1989).

7.7. Gene disruption

DNA introduced by transformation can cause disruption of the plant gene it is inserted into or a gene close to it (Feldmann *et al.*, 1989). The integration of the marker gene into genes essential for gamete or embryo development is more likely to give an obvious change in a plant's character as it is then clearly visible.

7.8 *Trans* gene interaction

Evidence from a range of organisms indicates that the presence of multiple copies of homologous DNA sequences in a genome can, in certain cases, result in *trans* interaction which has negative effects on gene expression (Jorgensen, 1990).

Matzke *et al.* (1989) have observed the inactivation of bacterial nopaline synthase (*nos*) and kanamycin resistance (*kan*^{r}) genes carried by T-DNA (T-DNA-I) on introduction of a second T-DNA(T-DNA-II). T-DNA-II carried part of a *nos* gene and other sequences homologous to T-DNA-I. The phenotypes of the genes carried by T-DNA-II were not suppressed, *i.e.* suppression was not observed to be co-ordinated. Lack of expression of T–DNA-I was correlated with methylation of the DNA. This phenomenon is termed epistatic interaction.

In addition to epistasis, a second type of gene interaction, termed co-suppression has been reported by Napoli *et al.* (1990). Co-suppression refers to the co-ordinate inactivation of both an introduced and endogenous genes. In contrast to the epistatic T-DNA interactions which are irreversible in somatic cells of a given plant (Matzke *et al.*, 1989), co-suppression shows somatic reversibility : a plant can switch from producing totally white to wild-type purple flowers. Krol *et al.* (1990) have reported that the introduction of an extra copy of its own *chalcone synthase* or *dihydroflavonol-4-reductase* gene into petunia did not result in over-expression of the gene product but in fact resulted in a dramatic reduction in gene expression. Linn *et al.* (1990) have also reported that transgenic *petuni*a plants containing multiple copies of maize Al gene with CaMV 35S promoter had lower levels of expression and more extensive promoter methylation than was observed in plants containing single inserts of the same gene. Similar suppressive effects in transgenic *Arabidopsis* plants have also been correlated with multiple copies and increased methylation of the transgene (Scheid *et al.*, 1990).

Although the interactions described above have been observed in transgenic plants, Matzke and Matzke (1990) have suggested that similar interactions might also occur between endogenous homologous sequences. For example, the flower pigmentation pattern of a commercial variety of *petunia* is similar to that produced by co-suppression in transgenic plants perhaps reflecting a natural case of co-suppression (Krol *et al.*,). In addition, as Jorgensen (1990) has suggested, the cycling in activity of several different maize transposable elements (Dennis & Brettell, 1990). Might be due in part to homology-dependent interactions between multiple copies of the element.

CHAPTER II

GENE DELIVERY SYSTEMS

1. Polycation-mediated transformation

Polyethylene glycol is an extremely hydrophilic molecule and it is able to remove much of the "free water" within a solution; "free water" being those molecules able to interact with charged (usually ionic) molecules soluble in water (Draper *et al.*, 1986). Thus, under high PEG concentration, macromolecules such as DNA can no longer stay in solution and will precipitate. The protoplast membrane is normally negatively charged. By virtue of phosphate groups, DNA is also negatively charged and therefore charge repulsion effects limit interaction between DNA and protoplasts (Draper *et al.*, 1986). At very high concentration, PEG also seems to minimize charged repulsion effects, thus cell membranes can come into close contact to allow fusion. Polyethylene glycol has also been shown to be a strong stimulator of endocytosis in plant protoplasts, enabling the uptake of large particulate matter such as whole chloroplasts, liposomes, or bacteria. Thus, in DNA uptake experiments, the total effect of PEG is somewhat unclear.

Many other hydrophilic, long-chain polymers such as polyvinyl alcohol, long-chain cations such as poly-L-ornithine, poly-L-lysine and dextran sulphate, and high concentrations of divalent ions such as Zn^{2+} and Cu^{2+} have also been used to achieve high levels of DNA uptake (Draper *et al.*, 1986).

Transformation methods using the polycation "polybrene" (hexadimethrine bromide) to aid DNA transfer to plant cells are new, although similar techniques have been used for several years in the transformation of animal cells (Chaney *et al.*, 1986, 1988). The major advantages of the polybrene technique are the following :

(i) It is less toxic than other polycations

(ii) The high transformation efficiency allows very small quantities of plasmid DNA to be used, thus it is relatively easy to introduce only one of a sequence per genome. Even with quite high plasmid concentrations only one or a few copies of the plasmid are integrated and concatemers of integrated plasmid are rarely observed

(iii) Since no carrier DNA is used, the integration of plasmid DNA is direct into the host genome, and so should enable analysis of the sequences surrounding the site of integration

(iv) The transformation frequency appears to be independent of whether the plasmid is supercoiled or linear, or if the insert has been removed from the vector. Frequencies also appear linear over the measurable range

(v) The technique also works equally well for plasmids up to at least 45 Kb if transformation efficiencies are calculated on a molar basis

(vi) The polybrene method has worked for both stable and transient expression.

In the past ten years, it has been demonstrated that direct gene transfer in plants is an efficient method of transforming protoplasts of higher plants. Factors such as electroporation, carrier DNA, the composition of incubation media, PEG concentration (Spivak *et al.*, 1984; Negrutiu *et al.*, 1987), the nature of plasmid DNA and genotype (Kohler *et al.*, 1987) have been recognised as influencing the number of transformed colonies.

DNA transfer to plant cells can be aided by using polycations such as polyethylene glycol (PEG), poly-L-lysine, poly-L-ornithene or DEAE-dextran with or without metal ions such as Zn^{+2}, Li^{+}, Cs^{+}, Rb^{+}, K^{+}, Na^{+}, Ca^{+2}, or Mg^{+2} (Potrukus *et al.*, 1985b; Lorz *et al.*, 1985; Werr & Lorz, 1986; Uchimiya *et al.*, 1986; Cocking, 1987; Junker *et al.*, 1987; Junker, 1987).

PEG has been used extensively to promote cell-cell fusion in mammalian cells as well as in plant protoplasts (Kao & Michayluk, 1974; Davidson, 1977). It has also been used for the transfer of external naked DNA molecules into plant protoplasts (Krens *et al.*, 1982).

DNA-uptake may be enhanced by polycations as a result of the interaction of the polycations at one or more of the following levels (Hasnain *et al.*, 1980):

(i) Protection of the exogenous DNA against nuclease degradation.

(ii) Increase in the permeability of the plasma membrane.

It is still not known which mechanisms are involved in the uptake of nucleic acids through plant cell membranes. It may be that DNA enters into the protoplasts by an active process such as pinocytosis.

Pinocytosis, means "drinking by cells". It is .. .ually the term for a complex process whereby a cell can take in nutrients and other substances to which the cell membrane seems impermeable (Solomon, 1960). The name comes from the Greek words for "drink" and "cell".

Pinocytosis is thought to be the mechanism by which entry of such macromolecules, including hormones and antibodies, is made possible, since it is difficult to consider that the small pores in the cell membrane or the solution in the plasma membrane could possibly permit such macromolecules to enter freely. Pinocytosis might also be a means of entry for other large molecules, *e.g.* DNA, which the cell might incidently engulf along with the medium containing the pinocytosis-inciting particles or ions, possibly inducing transformation.

A simple model for PEG action

The mechanism of cross membrane transport is either an unspecific membrane process or it may be specifically mediated by permeases that are activated in the presence of PEG. It was found that DNA can enter the cell in the absence of PEG, however, the rate of gene transfer was higher in the presence of PEG, *i.e.* PEG is an activator for a specific membrane process.

A biological membrane is a co-operatively stabilized organization of lipids and proteins which contain dynamic, locally limited structural defects which are candidates for the onset of further PEG induced perturbations, leading to permeation sites for enhanced material exchange.

Carrol (1972) has suggested that there is a definite interaction between polycations (e.g., PEG) and nucleic acids. Our data suggests that this interaction has a specific order (Sawahel, unpublished data), *i.e.*, PEG (+) combines with the protoplast membranes (—). When DNA (—) is added an interaction takes place between the PEG (+) and DNA (—) forming a membrane (—)/PEG (+)/DNA (—) complex. It also confirms Antonelli and Stadler's observation (1989) that polycations probably enhance absorption of the transforming DNA to plasma membrane. The mechanism involved after that is not clear but our data indicate that DNA may enter into the protoplasts by a metabolic process such as pinocytosis since a much higher DNA uptake was observed to take place at higher temperatures (Sawahel, unpublished data) (Fig. 1).

2. Particle gun

The biolistic process represents a completely new approach to the problem of how to deliver DNA into intact cells and tissues. High velocity microprojectiles are used to carry DNA or other substances past cell walls and membranes (Sanford *et al.*, 1987). It has been described in different ways and has been referred to as the particles acceleration method, the bio-blaster method, the particle gun method, the microprojectile method, the gene gun method and because DNA is being "shot" into cells, it represents a type of biological ballistics, hence the term "biolistics" and tissues (Sanford, 1988).

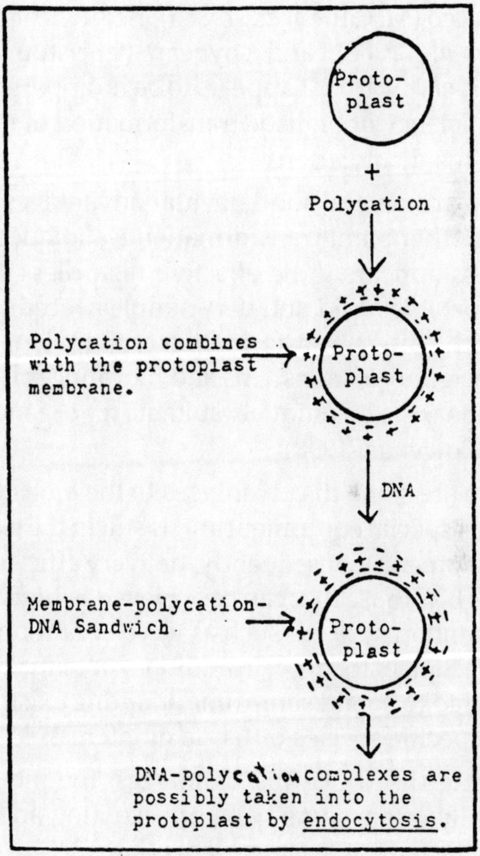

Fig. 1: A simple model for PEG action

The inventors of the process coined the term "biolistic" (biological ballistics) to describe both the process and any associated apparatus (Fig. 2) used to shoot biological material into living targets. A review of the literature reveals that DNA has been delivered and expressed in a number of plants including onion (Klein *et al.*, 1987), corn(Klein *et al.*, 1988 abc), tobacco (McCabe *et al.*, 1988a), rice, wheat and soybean (Wang *et al.*, 1988), and soybean (Christou *et al.*, 1988; McCabe *et al.*, 1988 b). It appears to be no apparent difference in the efficiency of biolistic transformation of mono-cotyledons versus dicotyledons.

There are several fundamental advantages of this process over other plant transformation techniques. The biolistic process appears to be effective regardless of species or tissue type; it is rapid and very simple procedure; it should facilitate the direct transformation of totipotent tissues such as pollen, embryo, meristems and morphogenic cell culture, and it appears to be uniquely suitable for organelle transformation (Potrykus, 1990).

There are some disadvantages to the biolistic process as it requires special equipment and is still in the early stages of its development. Consequently, delivery efficiencies are still not as high as those that can be achieved in highly optimized transformation systems such as electroporation or agrobacterial infection of tobacco. In addition, potential users should be prepared to spend some time adapting existing protocols to their specific species or tissue of interest.

Despite the above disadvantages, the biolistic process realistically offers some hope of becoming an "ideal" gene delivery system, at least in plants (Fig. 3).

The biolistic process employs high velocity microprojectiles to deliver DNA or other substances into intact cells and tissues. The process was first described by Sanford *et al.* (1987) and has been summarized in recent reviews by Sanford (1988, 1990a, b) and Sanford *et al.*, (1991). Evidence that the biolistic will eventually have a broad use has been

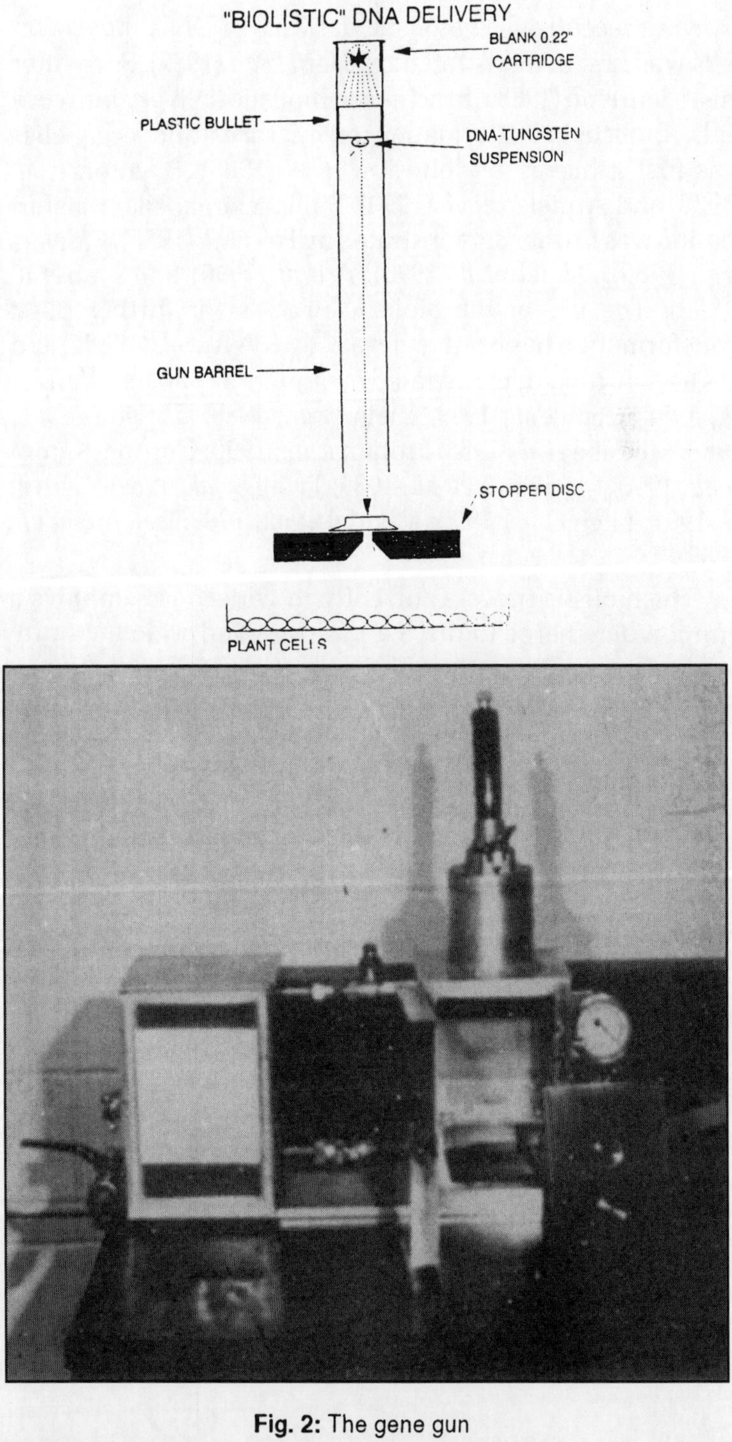

Fig. 2: The gene gun

growing recently. The biolistic delivery of DNA into onion cells was first demonstrated by Klein *et al.* (1987). Soon after this, Klein *et al.* (1988a, b) achieved biolistic delivery into corn cells. Biolistic transformation of microbes and organelles was first achieved by Johnston *et al.* (1988), Boynton *et al.* (1988) and Armeleo *et al.* (1990). Biolistic organelle transformation was further demonstrated by Fox *et al.* (1988), Blowers *et al.* (1989), Daniell *et al.* (1990), Ye *et al.* (1990), and Svab *et al.* (1990). The use of the biolistic process for further plant transformation has been reviewed by Klein *et al.* (1988c) and has been further demonstrated in numerous papers (Wang *et al.*, 1988; Cao *et al.*, 1989; Klein *et al.*, 1989; Christou *et al.*, 1988; McCabe *et al.*, 1988; Fromm *et al.*, 1990; Gordon-Kamm *et al.*, 1990; Morikawa *et al.*, 1989; Twell *et al.*, 1989; Klein *et al.* 1991; Fitch *et al.*, 1990; Mendel *et al.*, 1989; Zelenin *et al.*, 1989).

The biolistic process currently in widest use employs a gunpowder charge to drive a plastic cylindrical macroprojectile (Fig. 4) (Sanford, 1987; Klein *et al.*, 1987). Numerous laboratories have built variations of this system *e.g.* Zelenin *et al.* (1989). Morikawa *et al.*, (1989) built a device similar to the original macroprojectile device, but with the difference that pressurized nitrogen was used to drive the macroprojectile. However, with the Morikawa's device the macroprojectile velocities were distinctly lower than those of gunpowder-driven system. Likewise, Oard *et al.* (1990) have modified the original design by using a low-mass cylindrical macroprojectile drive by compressed air. McCabe *et al.* (1988) have built a macroprojectile accelerator where in the macro- projectile is a thin disk that is accelerated by the hot gases produced by a high voltage discharge, which vaporizes a droplet of water.

Different biolistic acceleration mechanisms are described by Sanford *et al.* (1987). These mechanisms include gas entrainment, transferred impulse, macroprojectile acceleration, centripetal acceleration and electrostatic acceleration

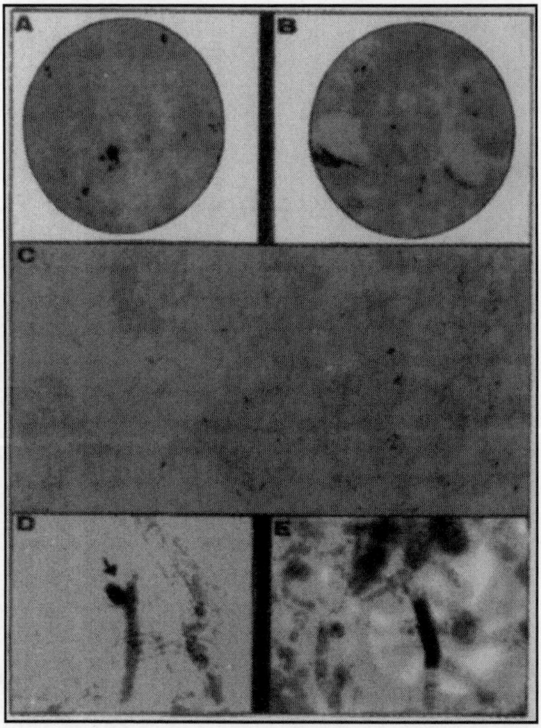

Fig. 3: *GUS* expression in *P. Patens* Protonemal tissue

A, B : Protonemal tissue before/after, bombardment

C : bombarded protonemal tissue showing a number of blue cells which are indicative of *GUS* expression

D : blue cells as neighbouring cells

E : blue cells as single isolated cells

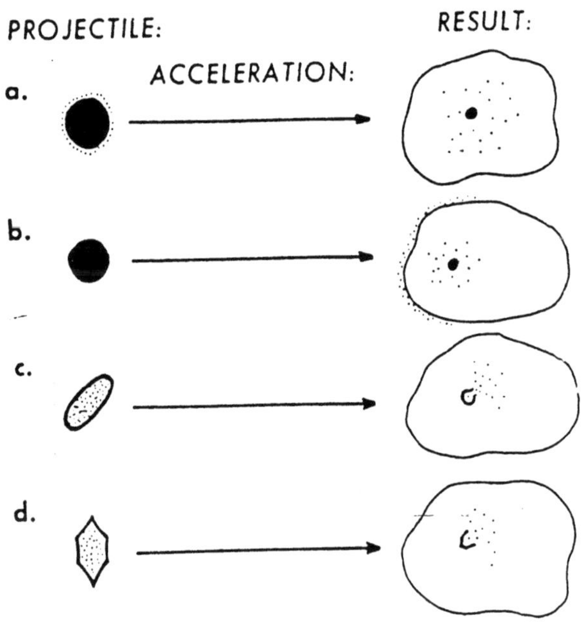

Fig. 4: Various embodiments of microprojectiles that can be used to deliver macromolecules into the interior of cells.

Macromolecules can either be on (a) or in the microprojectile (c or d). Macromolecules may also be carried into the cell in the wake of a penetrating microprojectile (b).

(Fig. 5). The acceleration of macroprojectile on the surface of a macroprojectile is the only acceleration mechanism at this time that has been proven to have wide utility.

In recent years, a large number of gene transfer techniques have been developed, including, *Agrobacterium* spp, electroporation, microinjection, *etc* (Sawahel & Cove, 1992). Although these techniques are useful for some plants, they are not universally applicable for a number of reasons which include a limited host range, low efficiency, or difficulty in regenerating plants from the transformed tissues that are produced. The use of microprojectiles as a DNA carriers (Klein *et al.*, 1989) has overcome some of the above obstacles. The technique appears to be adaptable to a wide variety of plant tissues, it can be used to treat large number of cells simultaneously and is applicable to intact, regenerable tissue.

Several factor might affect the efficiency of foreign gene delivery by this technique, namely, the physiological state and morphological character of the cells (cell wall thickness, size of cell aggregate, *etc.*); the DNA- carrying ability of the tungsten particles; the effective damage and; the degree of expression of the foreign gene in the bombarded cells. Although it is relatively easy to calculate the transformation efficiency (trans./μg DNA), the exact transformation frequency (trans./survivor) is extremely difficult to calculate due to the following factors :

(i) An uneven distribution of the particle over the plated cells.

(ii) The inability to obtain monolayers of plated cells (tissue composed of cell aggregates).

(iii) The difficulties encountered when counting the actual number of cells in the tissue before bombardment.

Why particle bombardment yield higher transformation efficiency than PEG procedure?

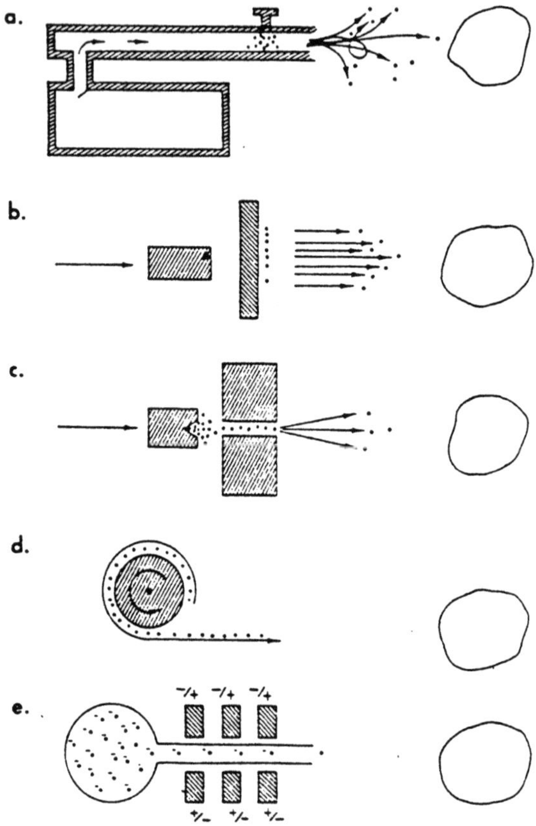

Fig. 5. The acceleration mechanisms that can potentially be used in the particle bombardment process.

(a) Acceleration by compressed air.

(b) Acceleration by transferres impulse.

(c) Acceleration by macroprojectile.

(d) Acceleration by centripetal firce.

(e) Acceleration by an electric field.

The high transformation efficiency could be related to the following :

(i) Only some of the cells are in a "competent" state with a high capacity to express the gene transferred into them. Competence here is viewed as the level of metabolic activity or the cell cycle stage and not as a function of some specialized cells. This competence state could, for instance, be reached in cells with high DNA repair activity. The tungsten particles used for bombardment may induce DNA damage (Sodeinde & Kindle, 1993) thus possibly induce repair mechanisms, *i.e.* possibly increase the number of competent cells. However, an activation of repair process could results from maceration of plant tissue and from the PEG procedure itself (Kohler *et al.*, 1990).

(ii) This may be a consequence of introducing a large number of DNA molecules into each cell.

(iii) Particle bombardment may introduce DNA directly into the nucleus.

3. Electroporation

During the process of electroporation (Fig. 6), a living cell is subjected to a strong electrical field. Electroporation is most commonly used to transport biochemicals such as DNA into a cell through pores that are thought to be created in the cell's surface in response to the electric field. The process of transporting biochemicals into cells is not very well understood, even though it has been widely used by biologists for some time. Peculiarities of cell physiology and cell wall and membrane structure preclude the possibility of any particular protocol for transporting biochemicals into cells being universal. However, among the most university applicable techniques for transporting DNA is electroporation. Citations showing successful electroporation into bacterial, yeast, fungal, mammalian and plant cells are

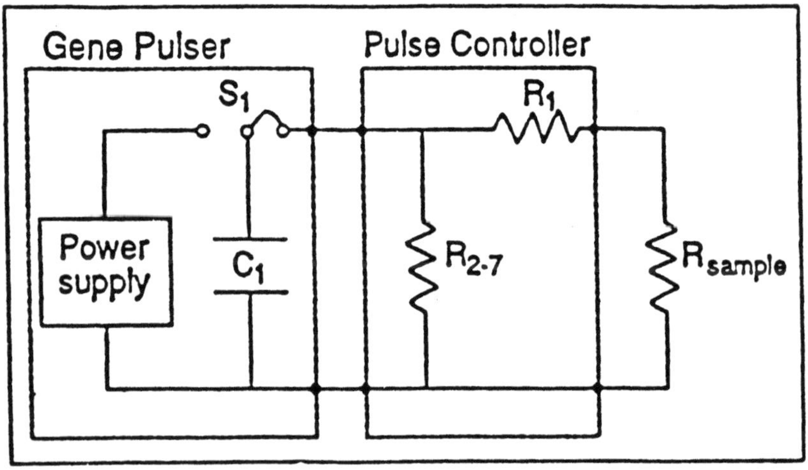

Fig. 6: Electrical circuit for electroporation

proliferating in the literature (Becker & Guarente, 1990; Potter, 1988; Simon & McEntee, 1990; Ward *et al.*, 1990; Meilhoc *et al.*, 1990; Rech *et al.*, 1990).

The advantages of electroporation include the following; it is a convenient, simple, fast and efficient method applicable to a wide range of plant protoplasts and involves low cell toxicity (Paszkowski *et al.*, 1984; Potrykus *et al.*, 1985b; Lorz *et al.*, 1985).

The main disadvantage of electroporation is that the system requires protoplasts and therefore a system for regeneration of these protoplasts to calli or whole plants. Consequently, it may not be applicable to some plant systems. However, D'Halluin *et al.*, (1992) have recently developed an electroporation system using organized and easily regenable tissue such as immature maize embryos.

A simplified picture of the important variables in electroporation

(a) In electroporation, an initial voltage "v" is applied to a cell suspension placed between two electrodes separated by a distance "d". This generates an "optimal" or "critical" electrical field, E = v/d (volts/cm), that is applied for some optimal or critical time period.

(b) The most important phenomenon dictating an upper limit to the voltage and time during electroporation is the heat generated by electrical energy. The amount of heat generated is directly related to the square of the applied voltage and time. If the heat generated during electroporation increases the temperature of the cells beyond their tolerable limits, cell death will result. Consequently, the apparent efficiency of electroporation diminishes if cell viability is essential to a desirable results. Clearly, excessive voltage is far more damaging to the cells than excessive time, since heat inactivation of the cells is related to the square of the voltage used.

(c) A second phenomenon that probably influences the efficiency of electroporation is electrophoresis in its most common application of electroporated DNA into cells. Both DNA and cells have a negative charge and migrate in an electric field in the same direction. The amount of electrophoresis that occurs during the electroporation process is directly related to the product of voltage and time. Electrophoresis results in an increased relative concentration of DNA and cells and thus, has a beneficial effect on electroporation efficiency.

(d) A general approach to electroporation designed to maximize beneficial effects and minimize detrimental effects requires minimizing the ionic content of the medium. In situations where ionic constituents are essential to a biologically appropriate environment, alternative non-ionic components in the medium (to osmotically support cells) might improve electroporation efficiency.

Mechanism

Neumann *et al.* (1982) have reported that the mechanism of cross membrane transport of DNA is either an unspecific membrane process or it may be specifically mediated by permeases that are activated in high electric field. This in turn may lead to a thining of a membrane patches and finally to a hole (Fig. 7).

4. Microinjection

The microinjection technique uses a fine capillary needle to deliver DNA precisely into cells (even nuclei of cells) in such a way that the cells are not killed but survive the treatment and can develop further into clones (Fig. 8).

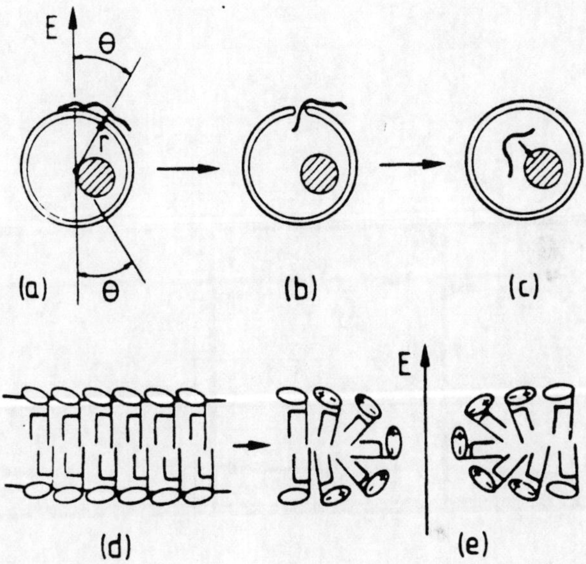

Fig. 7: Diagram of the electrically induced transfer of DNA into cells by the electroporation process.

(a) Model of a spherical biological cell with nucleus and an external DNA.

(b) Permeation of the DNA through a pathway induced by electroporation.

(c) DNA association with the nucleus.

(d) Unperturbed lipid domain in a membrane.

(e) Hole (pore), stabilized by a favourable lipid head group dipole interaction with the electric field.

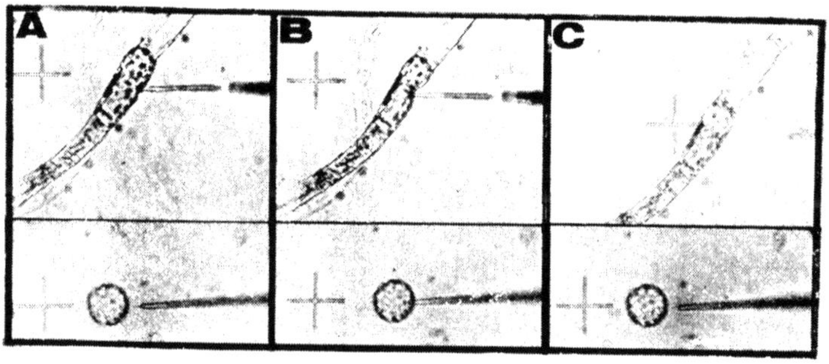

Fig. 8: Microinjection of cells and protoplasts

There are several advantages to the microinjection technique. First of all, the amount of DNA delivered per cell is not limited by the technique and can be optimized, thus improving the chance for integrative transformation. Secondly, the delivery is precise, even into the nuclei of the target cell, again increasing the chance for integrative transformation. Thirdly, the small structures with only a few cells and with high regeneration potential, for example microspore-derived and zygotic proembryos, can be injected. Finally, it is a direct physical approach and therefore host-range independent.

There are however some disadvantages, namely, that only one cell is targeted per injection and that the handling requires specialized skill and instrumentation. Whether the low number is a serious disadvantage is not yet clear because hundreds of injections can be performed per working day.

The microinjection technique, originally developed by both A. Graessmann and E. C. Diacumakos in 1970 for animal cells, has now become a routine method for the introduction of small molecules, macromolecules (DNA, RNA, protein) organelles and virus particles into a wide range of animal cells (Celis, 1984). The technique is based on the use of glass micropipettes with a 0.5-10 µm diameter tip to effect the direct transfer of macromolecules into the cytoplasm or the nucleus of a recipient animal cell or organ, immobilized on a solid support, artificially bound to a substrate, or held by a pipette under suction (Stacey, 1981; Ansorge, 1984; Hammer *et al.*, 1985)

Since microinjection of DNA had proven to be an effective technique for transforming animal cells, the development of comparable method for transforming plant cells was attempted by several groups (Griesbach, 1983; Steinbiss & Stable, 1983; Lawrence & Davies, 1985; Morikawa & Yamada, 1985).

Different application of microinjection technique in plant science have been reported. Some experiments dealt with the introduction of fluorescent dyes into differentiated cells to study intracellular transport (Goodwin & Erwee, 1985; Madore *et al.*, 1986). Efforts have been made to microinject microorganisms of virus particles into suspension cells (Nims *et al*, 1967 a, b; Halliwell & Gazaway, 1975; Bradley, 1979). Attempts to transfer chromosomes or even cell organelles have been made (Griesbach, 1987; Verhoeven & Blaas, 1988). One common feature of these experiments was that only very low numbers of cells could be manipulated due to the technical set-up used.

Microinjection apparatus

An inverted microscope (Nikon, TMD) equipped with epifluorescent optics is rested on a vibration-free table of a lamina flow chamber (Hitachi, PCHCS) (Fig. 9). A hydraulic microdrive (Narishige, MO-104) is positioned onto the microscope for micromanipulation of the injection micro-pipette. The injection micropipette (less than 1 μm outer diameter) is connected through a pressure regulator to a nitrogen bomb *via* copper tubing. The pressure regulator had two gauges, an high pressure gauge (up to 50 Kg/cm^2) and a low pressure gauge (up to 25 Kg/cm^2). The outlet gas pressure could be adjusted by switching a three-way stop-cock. By opening the outlet value, an appropriate pressure (usually less than 5 Kg/cm^2) could be applied to the injection pipette.

5. U.V. laser microbeam

The laser microbeam irradiation provides a new method of introducing exogenous gene materials into cells and has been successfully applied to tobacco (Weber *et al.*, 1992). DNA can be incorporated into whole plant cells and chloroplasts inside a cell by using a U.V. laser microbeam (Weber *et al.*, 1987). The basic idea of the laser method of DNA transformation is to let DNA contained in the culture medium enters the cell through a self-healing hole made with a laser microbeam (Tsukakaosh *et al.*, 1984).

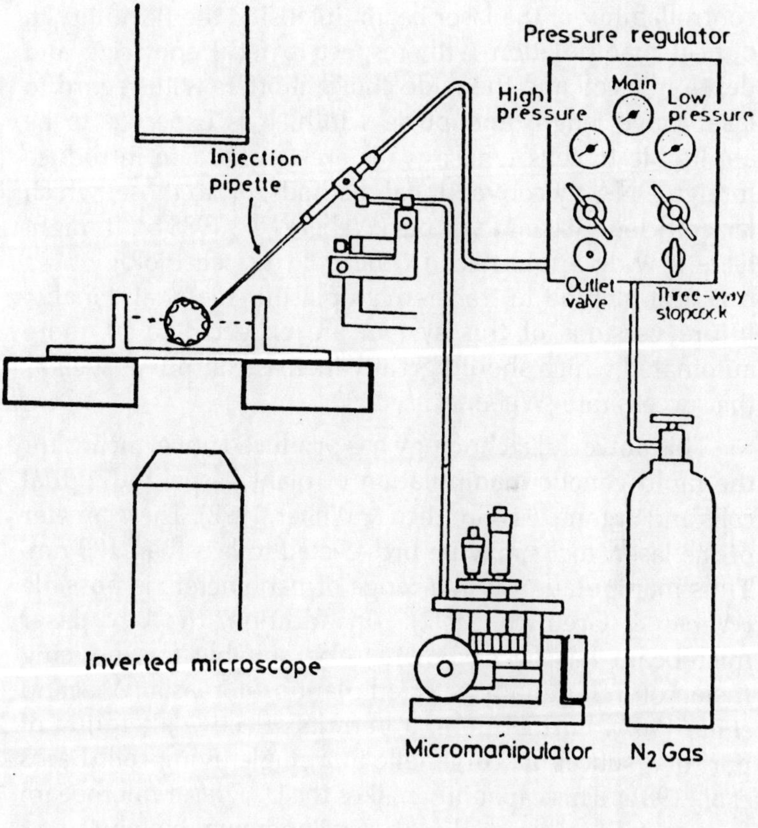

Fig. 9: Microinjection *system*

There are several advantages to this method. Unlike mechanical or electrical devices for transformation, a laser beam can be used inside of tissues, cells and organelles. It offers a clear advantage over the conventional methods both in the rate of transformation and in the speed of operation (Weber *et al.*, 1988a). It is extremely versatile due to the wide controllability of the laser beam intensity, the flexibility in optical manipulation with respect to focal spot size and depth in a cell and the wide choice it offers with regard to laser wave length and pulse width. It is expected to be applicable to cells whose walls are too hard to introduce foreign DNA by conventional methods, or, to those, which lack efficient plasmid vectors (Weber *et al.*, 1988 b). It might have a wide application in the field of cell biology as a practical method to transfer cytoplasmic material. Finally, future versions of this system are expected to be more automated which should greatly increase output as well as the success rate (Weber *et al.*, 1990).

This powerful technology has produced novel means for the rapid genetic manipulation of plant tissue, individual cells and organelles (Greulich & Weber, 1992). The diameter of the laser focus may be preselected to less than 100 nm. Thus manipulation in the range of nanometers is possible (Weber & Greulich, 1992). In addition, the U.V. laser microbeam has been shown to be suitable for inducing fusion of individually selected plant protoplasts (Wiegand *et al.*, 1987). Furthermore, it permits selective alteration of part of a subcellular organelle in a single living cell (Berns *et al.*, 1981). This capability makes the U.V. laser microbeam a powerful tool for studies in developmental biology.

The U.V. laser microbeam apparatus

The laser microbeam apparatus (Fig. 10) was described by Greulich and Weber (1992). An excimer laser (Lambda Physik, EMG 102 MSC) emitting 17-ns pulses at 308 nm with a repetition rate of 1— 100 Hz pumps a dye laser

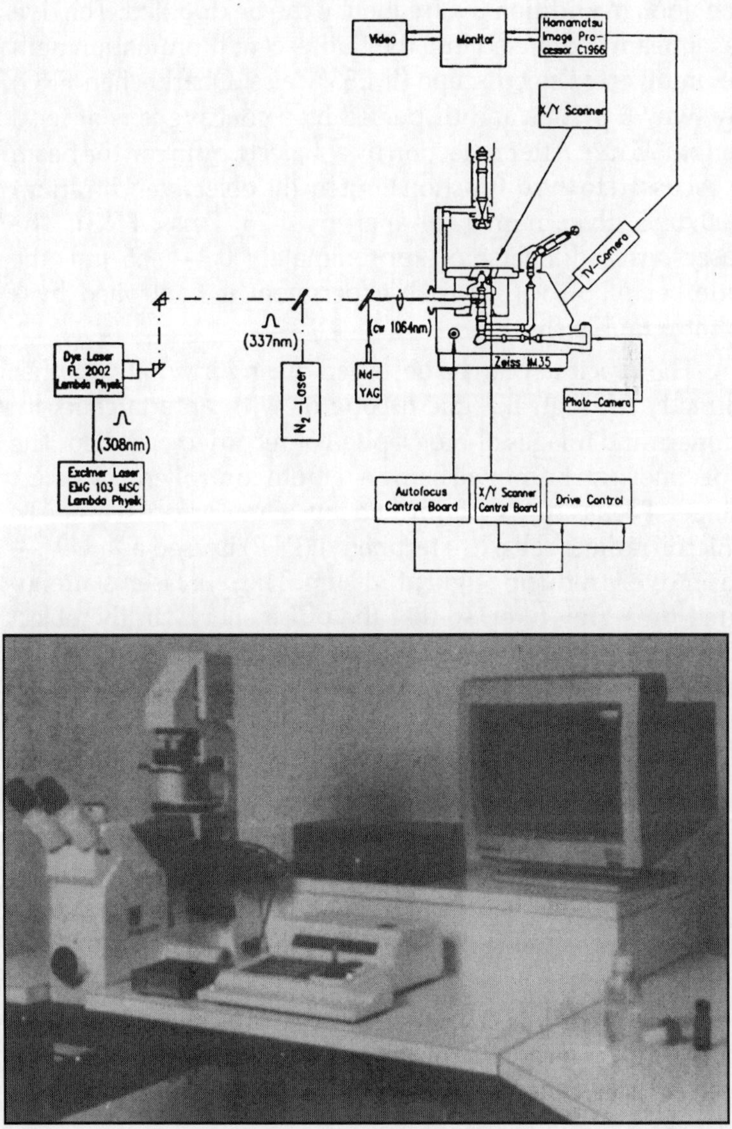

Fig. 10: U.V. laser micro-beam system

(Lambada Physik, FL 2002). The laser is tunable from 320-970 µm and , in addition the frequency can be doubled. The dye laser beam is directed into the fluorescent illumination path of an inverted microscope (IM 35, Zeiss, Oberkochen FRG) by two 90 prisms and dispersed by a concave lens of focal length 75 nm. After reflection by a dielectric mirror the beam is focussed to the diffraction limit by the objective (Ultrafluar 100/0.85 glyc., numerical aperture 1.25, Zeiss, FRG). The laser has a diameter of approximately 0.3— 0.5 µm. the number of pulses for each experiment is controlled by a photographic shutter.

The specimen can be observed during irradiation either directly through the microscope or with an attatched TV camera and images photographed or recorded on video. The specimen can be moved using a remote controlled x/y object stage. Focussing is performed by a motorfocus unit. To microirradiate cells, P— terphenyl (PTP) is used as a U.V.— laser dye, producing light at 340 µm. The dye laser is attenuated by a grey filter, so that the pulse energy in the object plane is adjustable between 1 and 10 µJ for all the experiments.

Direct DNA transfer

The present method employs ultraviolet laser light of nanosecond pulse length as an "optical" needle to prick cells (Tsukakoshi *et al.*, 1984). It perforates a self-healing hole of sub-micrometer in the cell membrane, through which the foreign DNA, contained in the medium, has a chance to slip into the cells.

The tissue is placed onto the microscope stage for treatment. A u.v. laser microbeam of 343 nm wavelength is directed at the cell wall and the plasma membrane of selected cells. A tunable dye laser is pumped by an excimer laser delivering pulses of 15 ns and an energy of a few mJ which make a hole in the cell wall. (Fig. 11).

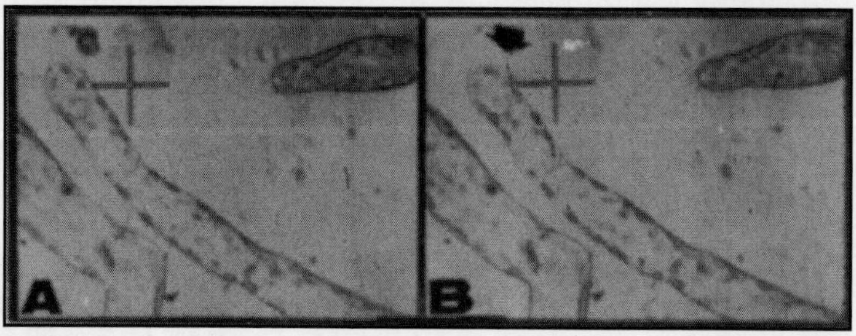

Fig. 11: U.V. laser-induced direct gene transfer

6. Electroinjection

Electroinjection (Fig. 12) is a direct transformation method for introducing genetic material into intact walled cells. This has the advantage that the cells totipotency is kept intact (Lindsey & Jones, 1990; Morikawa *et al.*, 1986; Nishiguchi *et al.*, 1987; Kohler *et al.*, 1987; Morikawa *et al.*, 1986).

The development of such a method has been vital for plant genetic engineering because in some plant species the ability of the plant cells to regenerate whole plant (totipotency) is enormously reduced when cell walls are digested to liberate protoplasts.

Electroinjection has none of the difficulties inherent in protoplast transformation and is, in principle, applicable to all plant species because as a physical method it is free of "host specificity".

7. Electrophoresis

Attempts to introduce DNA by electrophoresis into germinating seeds of barley (Ahokas, 1989) and to calli and seeds of various other species with a suitable embryo have so far proved unsuccessful. However, as yet, it is still not possible to decide just how valuable electrophoresis is as a technique.

8. Protoplast Fusion

The most direct method of gene transfer to plant cells is the uptake of plasmid DNA but at present its application relies on the availability of isolated genes. Until now, few genes for agronomic traits have been isolated and, in most cases, these traits are not coded for by single genes but rather by families of interacting genes. Thus, there is an interest in techniques which allow the transfer of large blocks of intact DNA encoding polygenic traits.

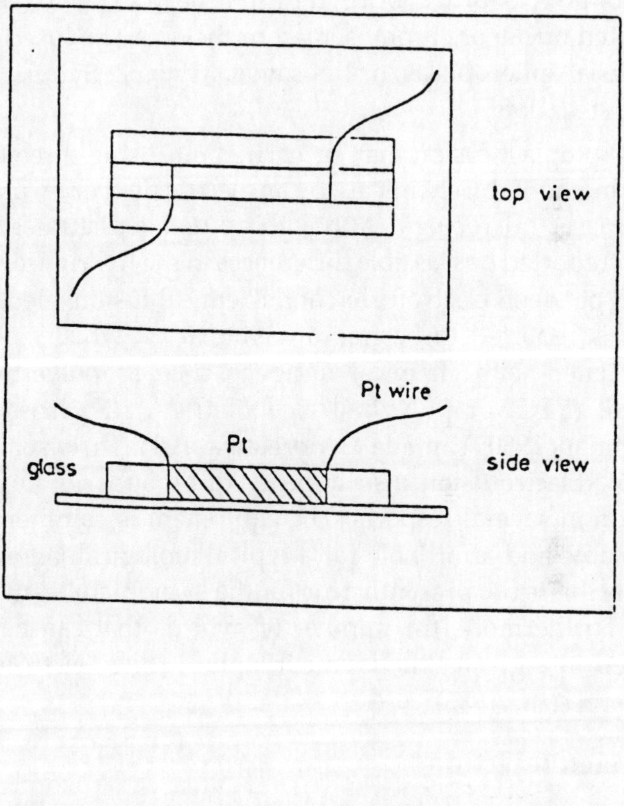

Fig. 12. Diagram of parallel electrode chamber.

Two platinum plates, which served as the electrode, were glued to four pieces of glass on a glass plate to form parallel electrode chambers about 2 mm wide, 2 mm high and 10 mm long (each chamber having a volume of about 40 μl). A piece of platinum wire was fused to each of the electrodes and connected to an electric circuit for the impulse generation. The chamber was placed in a petri dish (90mm) and autoclaved before use.

Protoplast membranes may be induced to take up particles by endocytosis or to fuse with other membranes thus making several of the above mentioned techniques possible. Protoplasts may be fused with other protoplasts to form somatic hybrids or with nucleated or mitotically inactivated protoplasts to form cybrids; they may be induced to take up isolated nuclei or chromosomes; or they may be fused with bacterial spheroplasts, or liposomes (Gleba & Sytnik, 1984; Lorz *et al.*, 1985).

Protoplast fusion may be carried out using three different methods; by chemicals, by an electric field or by using a U.V. laser microbeam. Although no detailed studies have been reported on possible differences in the hybrid isolation yield between electrofusion and chemical fusion, electrofusion is reported to be free of cytotoxicity, which is often associated with chemical fusagena such as polyethylene glycol (PEG) or polyvinyl alcohol (PVA) (Zachrisson & Bornman, 1980; Yamada & Morikawa, 1985; Turkaspa *et al.*, 1986). Electrofusion also appears to be superior to laser fusion in several respects. The equipment is commercially available and affordable for a typical biological laboratory while this is not presently true for the laser microbeam (Fig. 13). Furthermore, the number of hybrids that can be produced in a given time is much higher using electrofusion (Austin *et al.*, 1985).

9. Macroinjection

The macroinjection technique involves the use of injection needles with diameters far wider than cell diameters for delivering DNA into wound sites within tissues. However, at the same time it destroys those cells which are in direct contact with the injected DNA (Coe & Sarkar, 1966; Sayfer, 1980; Mathias, 1987; Zhou *et al.*, 1983) and to integrate, the DNA must gain entry to intact cells adjacent to the wound site.

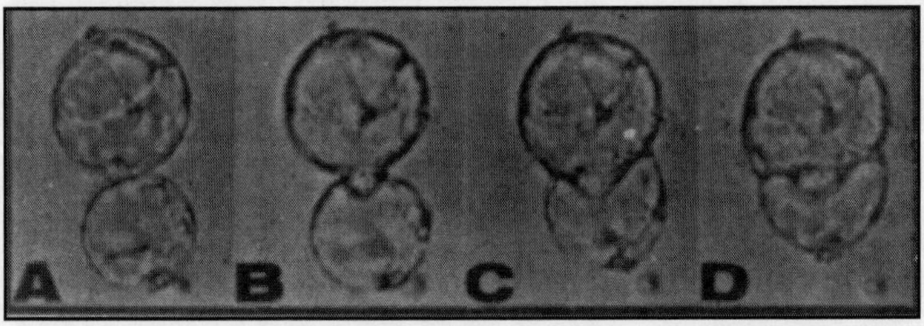

Fig. 13. Laser-induced protoplast fusion

There are a variety of problems associated with experiments that require DNA passage through one, two (in the case of pollen) or even many cell walls. First of all, the uptake of DNA across cell walls is a very rare event. Secondly, the majority of DNA molecules applied will attach to cell walls before they have a chance to enter the cell. Thirdly, each subsequent cell wall will reduce the chance of further penetration. Fourthly, DNA does not normally move from cell to cell. Finally, competent cells must be available; thus, even if DNA could occasionally cross one cell wall, the combination of several low frequency events will prevent the development of a routine and efficient gene transfer method (Potrykus, 1990).

A review of the literature reveals that little attention appears to have given to most macroinjection experiments into ovule, embryos and other tissues. However, one such experimental approach using immature floral meristems of rye (*Secale cereale*) fourteen days prior to meosis (De la pena *et al.*, 1987) appears to contradict all the arguments given above, and has yielded promising data, *i.e.*, foreign market genes detected by southern blotting and enzyme assay, indicate activity of the foreign gene in selected offspring. Whether or not this method can be applied to other species having gametangial development similar to rye has yet to be established, but such experiments are in progress in several laboratories.

10. Liposome system

DNA or RNA, encapsulated in liposomes, can be transferred into plant protoplasts by either direct fusion with the naked plasma membrane or through endocytosis of the liposome (Fraley *et al.*, 1980; 1982). Although it is efficient in terms of DNA encapsulation and transfer, the method once again requires protoplasts and its application is limited to species for which protocols are available for protoplast regeneration.

Foreign macromolecules, for example impermeant fluorescent dyes, RNA and DNA, can be introduced into walled plant cells by an alternative strategy which involves encapsulation of the desired probe into liposome that can be pressure-injected into the vacuole. There are two advantages of this method. First of all, the liposome protects the DNA from being damaged by the acidic pH and proteases present in the vacuole. Secondly, when liposomes are constructed from material that is compatible with the lipids of the tonoplast, liposome-tonoplast fusion occurs (Madore *et al.*, 1986) and so the encapsulated DNA should be successfully delivered into the cytoplasm (Lucas *et al.*, 1990).

Liposomes would appear to offer the following advantages as vectors for gene delivery into plant cells (Gad *et al.*, 1990):

(a) enhanced delivery of encapsulated DNA by membrane fusion.

(b) protection of nucleic acids from nuclease activity.

(c) delivery into a variety of cell types besides protoplasts by entry through plasmodesmata.

(d) delivery of intact small organelles.

11. Ca-DNA co-precipitation method

The basis of the Ca-DNA technique is the formation of a co-precipitate of plasmid DNA with calcium phosphate, formed by adding $CaCl_2$ to a solution of DNA in phosphate buffer. On contact with protoplasts, the co-precipitate is transferred across the plasmamembrane in a calcium-requiring process; and in order to encourage the endocytotic uptake of the precipitate, the protoplast/Ca-DNA complex is treated with polyvinyl alcohol and high pH calcium. This method was first used for plant transformation by Hain *et al.* (1985) using a tobacco leaf protoplast system.

12. Silicon carbide fiber-vortex

Recently, Kaeppler *et al.* (1990) developed a procedure for the injection of *Zea mays* and *Nicotiana tabacum* tissue cultures using silicon carbide fibres. This method involves vortexing a mixture of plasmid DNA encoding a selectable or screenable marker gene, silicon carbide fibers, and tissue of the species to be transformed in an Eppendorf tube. The silicon carbide fibers apparently act as microinjection needles and facilitate DNA delivery into the suspension culture cells.

Advantages of the silicon carbide fiber-mediated transformation method over other procedures includes the ability to transform walled cells (thus avoiding protoplast isolation), relative ease of the procedure, and very low equipment costs (Kaeppler *et al.*, 1992).

Mechanism

DNA delivery in this system is presumably due to cell-wall penetration by DNA-coated silicon carbide fibers. The silicon carbide fibers apparently act as microinjection needles and facilitate DNA delivery into plant cells. Cellular penetration is likely enhanced by frequent and forceful intercellular collisions encountered in the vortexed milieu (Kaeppler *et al.*, 1990). DNA bound to the fiber surface is carried into the penetrated cell and may have the potential to become integrated into the nuclear genome and stably transform plant cells in a manner similar to that proposed for the transformation of animal cells with asbestos (Appel *et al.*, 1988).

13. Sonication

The latest development involves use of ultrasound to facilitate the uptake of nucleic acids into plant cells (Joersbo & Brunstedt, 1992). Apparently, this technique provides an attractive alternative to traditional methods, particularly because plant protoplasts (Joersbo & Brunstedt, 1990), cells in suspension (Joersbo, 1990) and intact pieces of plant tissue (Zhang *et al.*, 1991) can readily be transformed.

Methods for ultrasound transformation

Gene transfer by ultrasonication employs the simple procedure irrespective of the nature of the plant material to be transformed.

The protoplasts, suspension cells or small pieces of tissue are suspended in a few ml of sonication medium in a microcentrifuge tube. Plasmid DNA (and possibly carrier DNA) is then added and after rapid mixing the samples are ready for sonication.

The pulses of ultrasound are delivered by ordinary machines (*e.g.* Branson, Sonics & Materials) used for homogenization of various tissues. The microtip (normally tapered) is immersed 2-3 mm into the suspension and pulses of selected intensity and duration are delivered. The cells are finally transferred to fresh growth medium.

Mechanism

At present the mechanism for acoustic permeabilization is not understood. However, two possibilities should be considered. The first regards the violent collapse of activation bubbles, generating high-pressure and high-temperature shock waves which could possibly cause localized rupture of the plasmalemma and lead to uptake of exogenous solutes followed by re-establishment of membrane integrity. The second hypothetical possibility originates from the electro-mechanical model presented by Zimmermann *et al.* (1974) predicting the existance of a critical hydrostatic pressure at which the intrinsic membrane potential is sufficiently high to induce mechanical breakdown of the membrane (Joersbo & Brunstedt, 1991). Reversible mechanical breakdown has been observed in human erythrocytes subjected to high pressures (60-100 MPa) in a hyperbaric chamber (Zimmermann *et al.*, 1980). Consequently, it is possible that either the high oscillating pressure generated by the ultrasonic field and/or the high-pressure shock waves originating from collapse cavitation could produce such higher hydrostatic pressures that reversible membrane breakdown would occur (Joersbo & Brunstedt, 1992).

Direct gene transfer

Ultrasonication has been reported to mediate gene uptake in plant protoplasts, suspension cells and intact pieces of tissue (Joersbo & Brunstedt, 1992).

a—Plant protoplasts

Joersbo and Brunstedt (1990a) found that plasmid DNA could be introduced efficiently into sugar beet and tobacco protoplasts by a brief exposure (500-900 ms) to 20 kHz ultrasound at 0.5-1.5 W cm^{-2} of acoustic power. Successful transformation was evidenced by transient expression of the introduced gene for chloramphenicol acetyltransferase. Optimal transient expression was obtained at rather high plasmid concentrations (40-100mg/1) and sucrose concentrations (21-28%) in the sonication medium. In sugar beet protoplasts transient expression was reported to be 7-15 fold higher than the expression obtained by electroporation (Joersbo & Brunstedt, 1990d).

b—Plant suspension cells

Sonication has also been shown to be able to facilitate uptake of plasmid DNA into intact suspension cells of sugar beet and tobacco (Joersbo, 1990a). Ultrasound intensity and duration were 2-3 fold higher than for protoplast transformation. As a result of sonication the majority of the cell aggregates were dissociated into single cells or small clusters of 1-5 cells providing a quasi-single system.

The mechanism for DNA uptake in sonicated cells remains to be established but formation of small breaks in the cell wall could possibly be involved in concert with the plasma membrane alterations hypothesized above.

c—Plant tissue

Zhang *et al.* (1991) reported stable transformation of tobacco by sonicating small pieces (4 x 8 mm) of leaf tissue. Sonication was performed at 0.5 W cm^{-2} for 30 min. which was approximately the same intensity as used for protoplasts, but the exposure time was 1500-2000 fold longer. Addition of carrier DNA and dimethylsulphoxide increased the transformation rate significantly. The transferred genes were inherited in a Mendelian fashion.

Comparison with other gene transfer systems

At present, a number of techniques exist for transferring genes into plant protoplasts and cells. Some of these techniques are confined to transformation of protoplasts because the cell wall, which is largely impossible to plasmid DNA, is generally not sufficiently permeabilized to permit uptake of DNA. As protoplasts of many species are often difficult to regenerate efficiently into fertile plants, the major asset of these techniques is an analytical tool, for example, in testing the function of gene constructs. Other direct gene transfer systems (*e.g.* particle bombardment) are designed for transformation of intact tissues.

However, most of the systems discussed require sophisticated and costly equipment. In addition, it does not allow the combination of both accurate quantitative estimates of gene expression and efficient generation of transgenic plants. Ultrasound transformation is unique among existing methods because of its versatility: both plant protoplasts, cell in suspension and intact pieces of plant tissue can be readily transformed. This feature, alongwith the fact that the equipment for ultrasound transformation is simple, cheap and multifunctional should allow for substantial future applications.

CHAPTER III

STRATEGIES FOR IMPROVING TRANSFORMATION EFFICIENCY

1. Plasmid DNA

(i) The type of plasmid DNA

Differences in the transformation frequency were reported when different plasmids were used. Kohler *et al.* (1987) have explained this phenomenon in two ways. First of all, some plasmids have a better uptake by protoplasts in comparison to others, possibly as a consequence of different structure and / or size of the plasmid. Secondly, the expression signals possess a distinct influence upon the transformation rates. This may be the consequence of different integration and / or expression rates. In addition, Negrutiu *et al.*, (1990) have reported that there are differences in the transformation rates of two plasmids containing the same gene expression signal. This might be due to differences in the promoter fragments, used length and nucleotide sequence of the untranslated leader or the untranslated ends.

(ii) DNA topology

The effect of DNA topology upon transformation is still controversial. Werr and Lorz (1986) have suggested that, compact circular DNA could be expected to be more effective than linear DNA for the entry of DNA into cells through the cell membranes. However, linear DNA may be more efficient for the expression of the introduced genes or integration of exogenous DNA. Ballas *et al.* (1988) have demonstrated that the higher expression observed with linear DNA takes place at the transcriptional level. This may suggest that, in certain cases, RNA polymerases of plant cells can interact more efficiently with linear than with circular forms of DNA, or linearization increases the efficieny of RNA termination processes and this in turn enhances the transcription efficiency. However, it is also possible that the expression observed with circular DNA is due to an activity of linear molecules formed intracellularly. Szostak *et al.* (1983) has suggested that circular DNA may be a poor substrate for integration compared to linear DNA and proposed that a free double stranded end is the initiator for recombination.

2. Carrier DNA

The effect of the presence of carrier DNA on the number of transformants recovered and the transformation frequency obtained was tested (Sawahel, unpublished data). It is likely that the higher numbers of transformants obtained in the presence of carrier DNA due to the fact that plasmids are frequently degraded and crumbled into smaller fragments which, in some cases, link with carrier DNA to form larger moleculed (called transgenomes) prior to integration (Jongsma *et al.*, 1987), *i.e.* carrier DNA is not neutral in function and may play an active role in stabilizing and fusing plasmid DNA sequences.

3. DNA repair

Experiments with animal, human and plant cells have indicated that the introduction of single-strand breaks, gap or additional DNA replication forkas into the chromosome, could increase the transformation efficiency by providing more gaps in the genomic DNA for plasmid integration. Postel (1985) ; Van Duin *et al.* (1985) ; and Gharti-Chhetri *et al.* (1990) reported the enhancement of the transformation efficiency in animal, human and plant cells respectively after the exposure of transformed cells to U.V. light or after treatment with 3-aminobenzamide.

There have been numerous attempts to explain the nature and the mechanism of the stimulatory effect of U.V. light on genetic transformation. Kohler *et al.* (1987) reported that DNA molecules may be linearized by irradiation resulting in a higher number of transformed colonies. This is in line with Shilliti *et al's* observation (1985) which stated that linear DNA is (2.5–10) fold more efficient than circular DNA when electroporation is used for direct gene transfer.

Ayad and Fox (1969) have demonstrated repair synthesis of DNA after an X-ray treatment of lympophoma cells with 1.5-4.0 GY. ^{3}H–labelled DNA uptake and integration was higher in irradiated cells. They have presented adequate evidence to show that repair processes were implicated in the integration of exogenous DNA. Spviak *et al.* (1984) suggested, on the basis of the enhancement of transformation observed in human cells that U.V. damage might play a role in stimulating the recombination of the plasmid and the recipient cell DNA. In addition, Kohler *et al.* (1989) speculated that DNA molecules, present in the nuclei at the moment or irradiation, are integrated into the genome to a large extent than in the situation of non-irradiated protoplasts, *i.e.* the recombination machinery of the cells is activated by irradiation.

Miller *et al.* (1983) found that DNA molecules, which are initially unlinked, are eventually integrated as a large concatenate into the host cell genome. This indicates that before or during incorporation, frequent ligation and recombinant events occur among transformed DNA molecules. Robins *et al.* (1981) showed that in stable transformants, the donor DNA is inserted at one (or a few) site (s) in the host chromosomal DNA, which suggests that the integration event might be one of the limiting steps in the transformation process. Since U.V. irradiation of cells enhance recombination (Dubbs *et al.*, 1974 & Hall *et al.*, 1980) it is possible that the U.V. stimulation of transformation is a consequence of the effect on recombination. In other words, U.V. light might stimulate the recombination of exogenous DNA molecules with (i) each other to form a concatenate and (ii) the host DNA for stable integration.

Van Duin *et al.* (1985) have suggested that the stimulation of the transformation is a property of lesions on which the excision repair pathway is operative. It is possible that the structure of lesions themselves (local deformation of the helix) produce stimulation of the transformation efficiency, by provoking recombination at those sites. Alternatively, U.V.-like lesions might induce some kind of S.O.S. response in the transformed cell, one of the consequences of which is enhancement of recombination and integration into the host genome.

Morgan and Cleaver (1983) reported that 3-aminobenzamide (3-BA) delays rather than prevents ligation of replication intermediates. Gharti-Chhetri *et al.* (1990) suspected that by slowing down the ligation of DNA replication fragments, 3-BA treatment would either facilitate the access of transforming DNA to DNA-free ends or some (error-prone) repair mechanism (s) would be activated (Negrutiu *et al.*, 1985).

4. Transformation of synchronized protoplasts

Efficient genetic transformation of plant protoplasts is a goal which might be brought nearer by a consideration of experiences with bacterial and mammalian systems (Lack, 1978). In the highly transformable bacteria, efficient DNA binding requires the development of a competent state, which arises at a specific stage of the bacterial culture growth cycle. Similarly, with mammalian cells in culture, it has been recognized that successful transformation requires that the cell population are cycling and the involvement of a specific cycle stage (Fig. 1) in "competence" or susceptibility has been suggested (Sompayrac & Danna, 1981). Wigler *et al.* (1979) have speculated on the existence of a subpopulation of "transformation competent" cells. They conclude that such competence is not a stably heritable trait but is, rather a "transient property" of cells. Recently, Meyer *et al.* (1988) and Okada *et al.* (1989) have indicated that cells in the M-phase of

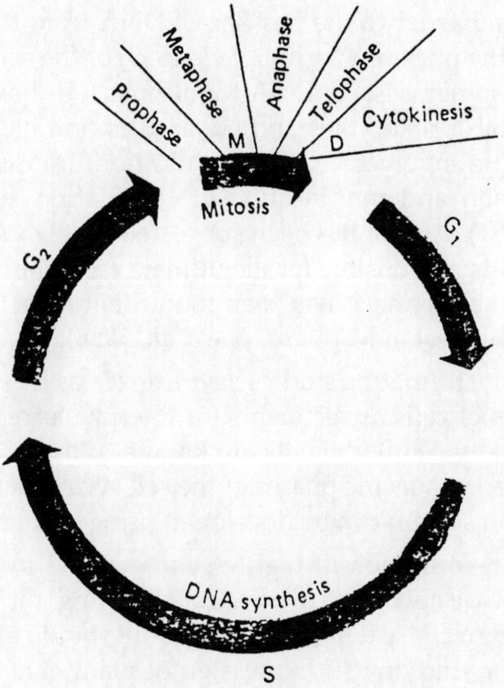

Fig. 1. The cell cycle.

The circle represents the cylindrical behaviour of regularly dividing cells, with the clockwise arrow representing the time direction; *i.e.,* the order of the successive phases in time. After cell division, there is a period during which no DNA replication occurs (the G1 phase). After this, DNA replication occurs during the S phase. Once DNA replication is complete, a phase (the G2 phase) precedes the condensation of the chromosomes at the beginning of mitosis. Mitosis, the M phase, is subdivided into the successive steps (prophase, metaphase, anaphase, telophase) which reflect the behaviour of the chromosomes. Nuclear and cell division then ensues (the D phase, which is relatively short) and the whole thing starts up again.

development are particularly susceptible for DNA transformation. The reason for this is unknown. It may be due to either the loss of nuclear membrane, at the M-phase, which constitutes a barrier to the passage of DNA from the cytoplasm into the nucleus (Yorifuji *et al.*, 1989) or the activity of DNA topoisomerases. As DNA topoisomerases have been found in both prokaryotes and eukaryotes and have been thought to be involved in DNA replication, transcription, recombination and modification of chromation structure (Gellert, 1981). Also, it has been suggested that DNA topoisomerase II is responsible for illegitimate recombination in mammalian cells and it has been found that its activity in cells is the highest at M-phase (Bae *et al.*, 1988).

Most transformation studies have utilized asynchronous populations of cells as recipients for the transferred DNA. Because, many cellular events are known to be induced in and confined to specific phases of the cell cycle, such a synchronization system can be developed using Aphidicolin.

The transformation of highly synchronized protoplast should provide not only a method for obtaining transformation by foreign DNA at high frequency, but should also offer a system for studying (i) Expression of plant genes which may be under the regulation of cell cycle such as those encoding histones (Heintz *et al.*, 1983). (ii) Integration modes. By detailed DNA analysis of transformants obtained from different cell cycle phase, it should be possible to differentiate between the integration modes.

5. Restriction-enzyme mediated event

Gharti-Chhetri *et al.* (1990) and Kohler *et al.* (1989) have reported that cellular DNA damage induced by U.V., X-ray, or ionizing radiation stimulates stable transformation by DNA-mediated gene transfer. However, DNA damage induced by these methods is too harmful to cells. Furthermore, DNA damage induced by X-ray or ionizing radiation is a mixture of double- and single-strand breaks and the results obtained by these methods are difficult to analyse (Yorifuji & Mikawa, 1990).

Recently, enzymatic restriction of mammalin cell DNA *in vivo* has attempted by cell permembilization with Sendai virus (Bryant, 1984), by simple agitation of cells in enzyme solution (Zhang & Dong, 1987) or by electroporation (Winegar *et al.*, 1987), and the resultant chromosomal abberrations or DNA-strand breaks have been reported.

Yorifuji and Mikawa (1990) have reported that co-transfer of plasmid DNA and restriction endonucleases, which will not cut within exogenous DNA, stimulates stable transformation of LtK cells by electroporation. They reported two possible explanations. One possibility is that DNA double-strand breaks in the cellular genome offer sites where exogenous DNA is inserted but the insertion is not simple ligation of cohesive ends as the stimulation was observed for various form of exogenous DNA. However, too many breaks as a result of a large amount of restriction enzyme may saturate the cellular machinery involved in the integration of exogenous DNA. Another possibility is that DNA double-strand breaks may mediate the integration of plasmid DNA into cellular genome by cleaving both plasmid and cellular DNA and religating the ends (Schiestl & Petes, 1991).

Schiestl and Petes (1991) have found that introducing the restriction enzyme *BamHI* together with a *BamHI*-cut DNA fragment increased the transformation efficiency of the yeast cells 7-fold higher than its absence. They also showed that the DNA fragment was often integrated into *BamHI* sites of the host genome. The simplest explanation of the restriction enzyme-mediated integration (REMI) is that the *BamHI* enzyme enters the cells, perhaps in association with the transforming DNA. The enzyme cleaves the chromosomal DNA at *BamHI* sites, the cohesive ends of the DNA fragments pair with those of the chromosomal DNA, and the resulting nicks are sealed by DNA ligasae (Schiestl & Petes, 1991).

Kuspa and Loomis (1992) have found that introducing any one of several different restriction enzymes alongwith linearized plasmid DNA stimulates the efficiency of transformation in *Dictyostelium discoideum* more than 20-fold over the rate seen when plasmid DNA alone is introduced. While the homologous enzyme, *BamHI*, stimulated transformation 27-fold, the heterologous enzyme *EcoRI* did not stimulate the rate of transformation above the level seen when no enzyme was added to the electroporation buffer. However, *Sau3AI* stimulated the rate of transformation as much or more than *BamHI*, considering that these two enzymes generate 5\ overhang (GATC). They concluded that for REMI to take place, the ends of the incoming plasmid DNA must match the recognition site of the restriction enzyme and argued in favour of the simple ligation repair model. They also found that REMI resulted in plasmid integration of the homologous restriction sites for the enzymes used in > 70 % of the transformants indicating that integration may be quite random.

Restriction enzyme-mediated events have potential uses such as generation of chromosome aberrations, stimulation of mitotic recombination at specific sites, and improvements in the efficiency of gene targeting in organisms with high levels of non-homologous integration (Schiestl & Petes, 1991). Also, Kuspa and Lommis (1992) have used REMI for insertional mutagenesis as the plasmid can integrate into broadly distributed sites throughout the genome, insertion might tag different genes and allow their rapid isolation.

6. Transformation booster sequence

Although DNA transfer systems mediated by *Agrobacterium tumefaciens* are very efficient (Schell, 1987), direct DNA transfer into plant cells has gained some importance since its first application (Davey *et al.*, 1980 ; Krens *et al.*, 1982). Several strategies to increase the relatively low transformation frequencies of direct DNA transfer have been tested. Considerable improvement was gained through transformation by electroporation (Shillito *et al.*, 1985) or using optimized Mg^{2+} and PEG concentrations (Negruti *et al.*, 1987). However, both these methods include the use of linearized plasmid DNA or of an excess of carrier DNA. Although transformation frequencies are remarkably high, the DNA integration pattern is complex (Shillito *et al.*, 1985).

The present advantages of *A. tumefaciens*-mediated transformaion are not only the high transformation frequencies obtained but, importantly, the simple integration pattern of the transferred DNA that guarantees practicability. Therefore, any technique for direct DNA transfer should improve the quality of the transformation products as well as the frequency of transformation. One way to achieve a more simple integration pattern is the use of supercoiled plasmid DNA without the addition of carrier DNA, since it has been shown that in animal system this leads to a greater proportion of undergraded and unrearranged integration events, while linear DNA often is subject to degradation (Gusew *et al.*, 1987). The use of circular DNA has the disadvantage of relatively low transformation frequencies (Krens *et al.*, 1982). This could be overcome when protoplasts were transformed had been synchronized into M phase (Meyer *et al.*, 1985). Relative transformation frequencies in the percent range were reached, while the simple integration pattern was conserved. Since synchronization procedures are not tolerated by all species and the technique is time consuming, alternative ways are required. Recently, Meyer *et al.*, (1988) reported that a 2-Kb genomic fragment of *Petunia hybrida* is responsible for at least 200-fold increase in the relatively low transformation frequencies that are achieved when supercoiled DNA is used for plant protoplast transformation. This sequence, which was named transformation booster sequence (TBS), shows its effect when inserted upstream of the selectable marker gene in either orientation. They also reported that the TBS effect was not caused by autonomous replication or improved stability of transforming DNA.

Through sequence analysis of TBS, an octamer GTGGTATG that corresponds to the core consensus sequence common to many transcriptional enhancer elements was found (Weiher *et al.*, 1983 ; Gillies *et al.*, 1983). Exchanges within this enhancer core region drastically redue transcription *in vivo*. These enhancer elements can activate transcription from promoters located either upstream or downstream; they are functional in both orientations and can be located > 1 Kb away from the promoter. The elements act in *cis* and often provide an open chromatin structure or a nucleosome-free gap (Schaffner, 1985). Another characteristic of many transcription enhancer elements are sets of repeated sequence motifs (Ebert *et al.*, 1987). Upstream of the octamer core consensus sequence in TBS, two sets of repeated sequence elements were also observed (Meyer *et al.*, 1988). The tetramer GCAG appears five times and a 16-nucleotide motif CAGACTCAGCTGACTC is imperfectly repeated twice.

The structural characteristics of TBS as well as its orientation independence suggest that TBS could also lead to a transcriptional enhancement of the integrated selectable marker. In the absence of TBS, the 35S promotor containing a transcription enhancer leads to 10-fold increase in transformation frequency. However, neither in transient assays nor in stable transformation did TBS have an influence on the strength of the *nos* promotor fused to the *NPTII* gene. Hence, the consensus core enhancer sequence observed on TBS is not active and that TBS seems to mediate its effect differently. It has been suggested that TBS either might guide the plasmid by an unknown mechanism to a chromosomal location from which transcription of the neighbouring *NPTII* gene is guaranteed or confer an open chromatin structure, which would facilitate transcriptional activity independently of the position of the vector DNA in the genome (Meyer *et al.*, 1988). This could lead to a higher proportion of integrated copies that would be expressed and therefore selectable.

CHAPTER IV

ORGANELLE TRANSFORMATION

Organelle transformation

The tools of molecular biology offer a plethora of ways to explore gene function and regulation, one of the most important being the manipulation of genomes by transformation. Integration of exogenous DNA into the nuclear genome has become routine for many plant species, and transgenic plants harboring, in their nuclear DNA, genes from other plants or from different organisms can be readily produced by a number of techniques. By contrast, techniques for the transformation of plastid genomes have been developed only recently (Fig. 1).

Transformation of plastid genomes has several potential problems that nuclear transformation does not share. For one thing, it may be more difficult for DNA to cross the plastid double membrane than the nuclear membrane (Maliga, 1993).

In addition, plastid genomes are present in much higher copy number than nuclear genomes—the single *Chlamydomonas* chloroplast has 80 genome copies, and higher plants can have 100 or more plastids, each with many genome copies. For a transformed genome to replace all copies of the original genome, therefore, strong selection pressure (for prototrophic growth or for antibiotic resistance) must be applied (Chasan, 1992).

Reports on transient gene expression suggest that there may be several ways to introduce DNA into the plastids of flowering plants, including *Agrobacterium* vectors (De Block *et al.*, 1985), polyethylene glycol (PEG) (Sporelein *et al.*, 1991), biolistic DNA delivery (Daniell *et al.*, 1990; Guang-Ning *et al.*, 1990; Daniell et al., 1991), and temporary holes cut by a U.V. laser microbeam (Weber *et al.*, 1991). Rigorous evidence for transient expression of the introduced transgene in plastids has been provided in two cases (De Block *et al.*, 1985; Sporelein *et al.*, 1991) ; however, one of these (De Block *et al.*, 1985) could not be reproduced. To date, only the biolistic

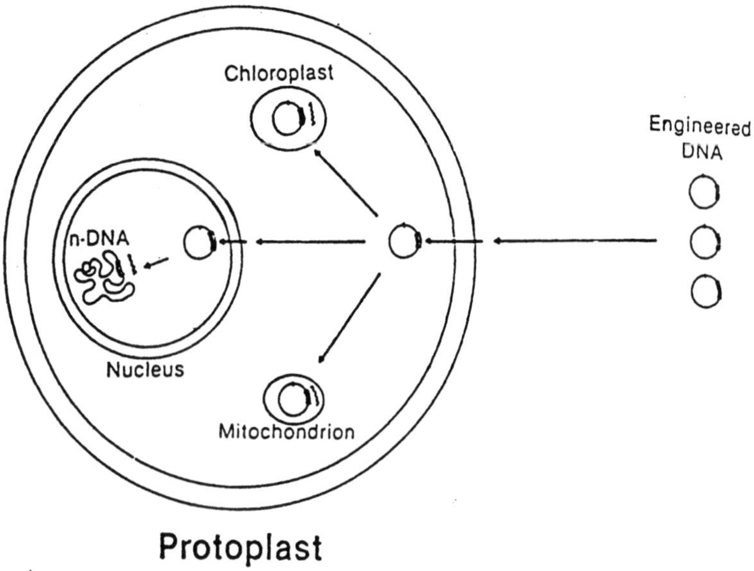

Fig. 1. Genetic transformation types

process (Svab *et al.*, 1990; Svab & Maliga, 1992, 1993; Svab & Maliga, 1993), and PEG treatment (Golds *et al.*, 1993) have yielded stable plastid transformants.

As the plastid transformation system has been developed for the model plant, tobacco, the technique will be applied initially to address fundamental questions of plastid biology including mechanisms of plastid gene regulation (Mullet, 1988 ; Gruissem, 1989), regulatory interactions of plastid nuclear genes (Taylor, 1989 ; Mayfield, 1990), editing (Hoch *et al.*, 1991) and translation (Bonham-Smith & Bourque, 1989) of mRNA, and photosynthesis (Rochaix, 1992).

CHAPTER V

SHOTGUN TRANSFORMATION

1. Plasmid rescue

Gene transfer into cells and organisms has been fundamental to molecular genetics and applications in biological research. In some situations it has proven desirable to retrieve the transferred DNA, either as a means to isolate that DNA after functional selection or to access the chromosomal DNA adjacent to the site into which the transferred DNA is integrated (Clark *et al.*, 1992).

One of the strategies that has been used for this purpose is plasmid rescue, whereby cyclized fragments of a cellular genome that includes integrated bacterial plasmid DNA are used to transform *E. coli.*, resulting in the rescue of the integrated plasmid and any cellular sequences to which it is linked.

Although plasmid rescue has been a powerful technique in both yeast and flies, where it has been widely employed for isolating genes interrupted by gene transfer (Winston *et al.*, 1983 ; Orr-Weaver *et al.*, 1983 ; Cooley *et al.*, 1988) the frequency of rescuing plasmids from mammalian cells has been dramatically below expectations (Tarantul *et al.*, 1986).

Plasmid rescue can provide an efficient technique for recovery of the tagged plant sequences (Koncz & Schell, 1986; Koncz *et al.*, 1989; Mayerhofer *et al.*, 1991; Behringer & Medford, 1992) and gene rescue after shotgun transformation by using either *Agrobacterium* transformation or total genomic DNA (Simoens *et al.*, 1986 ; Gallois *et al.*, 1992). However, cloning by rescue has been hampered by rearrangements in the rescued plant DNA sequences (Ott & Chua, 1990 ; Behringer & Medford, 1992).

Further Ott and Chua (1990) reported that the *E. coli* strain DH5α is capable of rearranging not only plant DNA but also sequences within the T-DNA itself (deletion derivatives accounted for 26 per cent of the rescued plasmid clones tested). Mandal *et al.* (1993) indicate that the observed rearrangements are caused by *E. coli's* ability to restrict c-methylated DNA (Noyer-Weidner *et al.*, 1986 ; Releigh & Wilsôn, 1986 ; Raleigh, 1992). This modified cytosine restriction (mcr) system is present in several commonly used *E. coli* strains including DH5α. By employing an *mcr*AB mutant of *E. coli* DH5α for plasmid rescue, Mandal *et al.* (1993) were able to avoid the rearrangements completely. All plasmids rescued from a particular transformed plant line were identical to each other and their structure corresponded exactly to that of the respective tagged genomic DNA in the plant (Mandal *et al.*, 1993).

DH10B as an ideal host strain for plasmid rescue.

E. coli is capable of degrading foreign DNA sequences that contain methylated cytosine and adenine residues. Restriction systems that function in this way have been identified and are designated *Mcr*A, *Mcr*B, *Mrr*. The *Mcr*A and *Mcr*B (modified cytosine restriction) systems restrict DNA that contains methylated cytosine residues, while the *Mrr* (methylated adenine recognition and restriction) system restricts DNA that contains methylated adenine residues (Raleigh & Wilson, 1986 ; Noyer-Weidner *et al.*, 1986; Heitman & Model, 1987).

The use of a host strain that does not restrict DNA at sequences containing methylated cytosine and adenine bases is clearly a requirement for the construction of representative genomic libraries, as well as for other procedures that involve the introduction of methylated DNA into *E. coli*, such as cDNA cloning (Raleigh *et al.*, 1988 ;) or plasmid rescue.

E. coli strain DH10B is *Mcr*A, *Mcr*B, *Mcr*C, and Mrr due to the introduction of an McrA mutation and the deletion of the entire *Mrr-bsd*RMS-*Mrc*BC region. The absence of these restriction systems allow DH10B to produce more colonies per μg of methylated DNA than other *E. coli* hosts. In addition to the markers that make DH10B an ideal host for cloning methylated DNA, the DH10B strain contains the following genetic markers : *rec*AI, which increases the stability of inserts ; *end*AI, which improves the quality of DNA from mini-preps; and *bsd*RMS, which allows cloning of DNA without cleavage by endogenous endonucleases. The DH10B strain also contains the ϕ80d*l*acZN15 marker that provides α—complementation of the β-galactosidase gene from pUC or similar vectors for blue/white screening on plates containing X-gal (Fig. 1)

2. Gene rescue

The availability of genetic transformation system opens up a number of areas to the transformation approach. The most important application is the possibility of isolating single genes from plant gene banks by complementation of mutant lines and the transfer of trait selectable in culture *via* transformation with cloned genes of total genomic DNA.

The ability to transfer, identify and possibly isolate genes encoded for traits which are monogenic and dominant (at least from certain plants) by the shotgun rescue system represents a qualitative technical advance that has created entirely new scientific and technological opportunities as many important genes are not amenable to traditional approaches which are used for gene isolation such as library screening with probes or antibodies.

The shotgun rescue system offers at least two strategies. The first one involves screening for a loss of function due to a gene disruption which can be caused by the insertion of a transposon (Coen *et al.*, 1990 ; Gierl & Saedler, 1992), or a T-DNA (Feldmann, 1989). We recently found that the frequency of DNA insertion mutagenesis increased upon the re-transformation of stable transgenic lines with related plasmids encoding different screenable and selectable markers which led to disruption of the functions of the endogenous genes controlling morphogenesis (Fig. 2) (Sawahel *et al.*, 1994). As only a few insertions are expected in a given line this approach necessitates the screening of a large population of individuals bearing insertions at

Fig. 1. Characterization of the transformed *E. coli* cells.
(A) The ability to grow under dominant selection.
(B) By using the *lac* Z gene as a genetic marker.

Fig. 2. GUS expression patterns in single & double transformants of the moss *Physcomitrella patens.*

A, B: Protonemal tissue and gametophore of single transformant.

C: Mutant double transformant.

different loci. A second strategy is to screen for a gain of function by the introduction of genomic DNA from a donor genome into a species lacking the trait of interest (Klee *et al.*, 1987 ; Simoens *et al.*, 1986). This strategy requires a high transformation efficiency. Model experiments for this second strategy have been carried out where genomic libraries have been transferred into plants via *Agrobacterium* with the view of recovering a gene of interest. These studies have shown that this strategy is technically possible. However, in practical terms, the general applicability of the approach may be limited due to the genome size, transformation frequency and ability to score for the presence of a gene in the transformant as (Simones *et al.* 1986) calculated that 17×10^6 transformants must be generated to transfer a particular DNA sequence from the tobacco genome into a host plant via *Agrobacterium* transformation.

Gallois *et al.* (1992) demonstrated that it is possible to identify a dominant plant gene encoded by 8 Kb or less by co-transferring the total genomic DNA and a plasmid carrying a selectable marker into a host plant cell using direct gene transfer (Fig. 3). Both the *Arabidopsis* and tobacco genomes were used as a DNA donor. The tobacco genome being 30 times bigger than the *Arabidopsis* genome. Also, they estimated that about 2000-3000 transformants would be enough to transfer a dominant gene with a tag linked to it from one given species to another. The above transformation frequency (even though its low compared to 17×10^6 using *Agroabacterium*-mediated transformation) would allow the application of a selection or screening scheme at the whole plant level to detect the transfer of a dominant trait from a donor species to a receptor. Perucho *et al.* (1980) has shown that ligation of a donor genomic DNA fragments to a linearized plasmid DNA prior to transformation (Fig. 3) allows the plasmid sequences to serve either as a physical or biological marker.

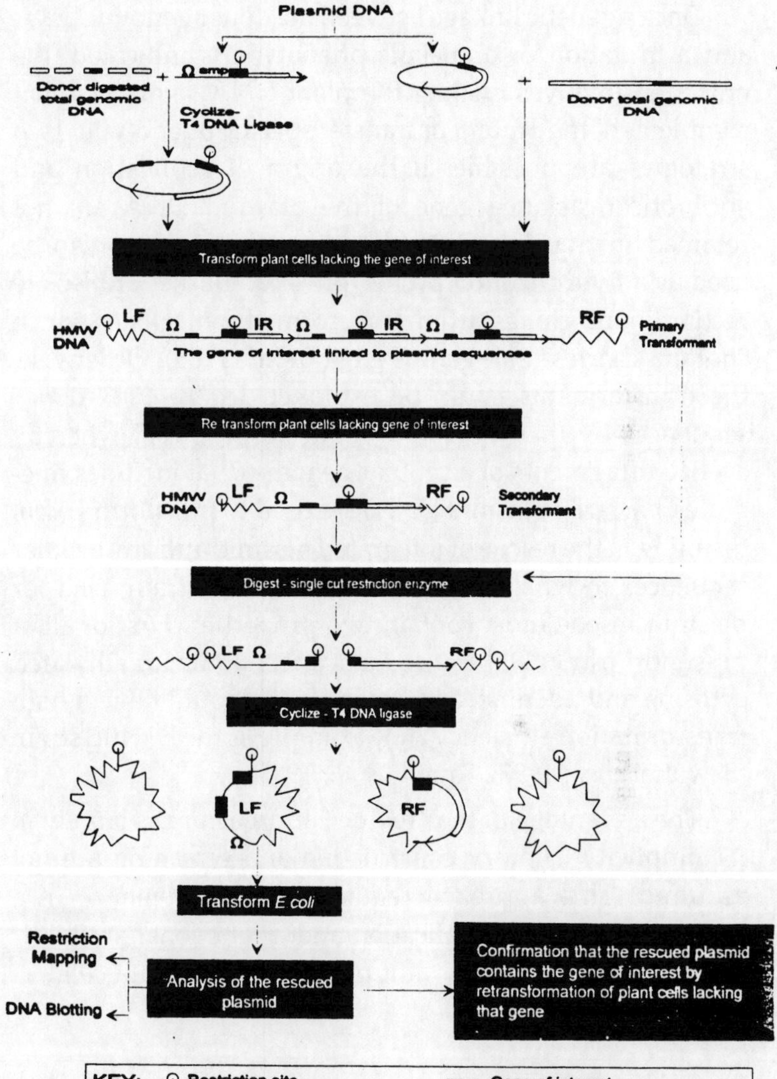

KEY:- ♀ Restriction site — Gene of interest

Ω Plasmid origin of replication and antibiotic resistance gene

LF/RF Left and right flank junctions between the chromosome and the transgenic array

IR Internal repeats of adjacent plasmids

HMW High molecular weight

Fig. 3. Strategy for gene rescue in plants

Once a genetic linkage between the donor genomic DNA and a mutation or a specific phenotype is indicated, the next step involves cloning the plant DNA sequences that complement the mutant or transfer specific phenotype. Two strategies are possible. If the origin of replication and antibiotic resistance gene of the plasmid DNA are not retained in transformants, the plasmid sequences can be used as a hybridization probe "physical marker" to isolate recombinant clones from transformed cell DNA which contains the desirable gene (Perucho *et al.*, 1980). Otherwise, the transformants could be exploited to "rescue" donor sequences by their ability to replicate in *E. coli*, whereby cyclized fragments of a cellular genome (that includes integrated bacterial plasmid DNA) are used to transform *E. coli*, resulting in the rescue of integrated plasmid and any cellular sequences to which it is linked The *E. coli* strain used for plasmid rescue must contain two properties. First of all, it must not restrict DNA at sequences containing methylated cytosine and adenine, and secondly, it should have a high transformation efficiency : one example, is the DH10B strain (Sawahel *et al.*, 1993 ; Knight *et al.*, 1993).

The advantage of the "rescue" approach rests largely in its simplicity : library construction and screening are not required. This approach requires the maintenance of a competent origin of replication and a functional antibiotic resistance gene through multiple passages in plant cell.

Prospects

Many interesting or useful genes that have been known for years on the basis of mutant phenotypes can now be isolated and characterized using the gene rescue system, *e.g.* single gene mutants tolerant to NaCl in the fern *Ceratoperis* (Warne & Hickok, 1987). In addition, the combination of gene rescue and insertional mutagenesis (Kuspa & Loomis,

1992) by restriction enzyme-mediated integration (REMI) of the plasmid DNA has allowed the tagging of developmental genes as well as cloning of the disrupted gene.

This ability to isolate genes for which mutations are available provides a substantial new motivation for expanding the collection of mutations that affect characters of scientific importance.

3. Promoter & enhancer rescue

This strategy relies on the activation of reporter genes, which possess either a minimal promoter or no promoter, by native gene regulatory sequences following transformation and random integration (Hope, 1991). Such activatable reporter sequences were described by Topping *et al.*(1991) as "interposons".

The rationale of the technique is that promoterless genes may be transcriptionally activated by integration downstream of a native gene promoter, to create either transcriptional or translational fusion [promoter trapping]; while genes with weak promoters may be activated by nearby native enhancer sequences, and will function in the absence of gene disruption [enhancer trapping] (Lindsey *et al.*, 1993). In both cases it has been reported that the pattern of reporter expression reflects the pattern of expression of the native gene (Kertbundit *et al.*, 1991). The isolation of the promoter sequences can be accomplished by cloning genomic DNA sequences in front of the promoterless reporter gene, followed by selection for those DNA fragments that activate the silent gene. Lindsey *et al.* (1993) have demonstrated the activation of an interposon in three plant species [*Nicotiana tabacum, Arabidopsis thaliana* and *Solanum tuberosum*] which have different genome sizes, different amounts of repetitive DNA and different ploidy levels.

The promoter and enhancer rescue approach can be used using the following experimental system:

(I) Genetic transformation, using interposons [*e.g.* A promoterless *uid*A (*gus*) gene, encoding the β-glucuronidase (GUS)reporter]

(II) Selection of the transformants.

(III) Detection of the transformants which have a fusion gene activity using histochemical GUS assay

(IV) Isolation of the genomic sequences flanking the interposon using Southern hybridization and plasmid rescue.

The genomic DNA should be digested and religated. The restriction enzymes used should be designed to generate interposon/moss DNA border fragments.

In order to usefully exploit this approach for studying patterns of gene expression during plant development, Lindsey *et al.*(1993) considered that it was important to characterise lines transformed with an interposon integrated at a single locus. They suggested that this would simplify the interpretation of expression patterns, which might be complex should transcription be initiated from two or more promoters,and would facilitate cloning of genomic sequence directing an expression pattern of interest.

The study of DNA sequences regulating gene expression in a specific way is crucial for the understanding of the morphogenetic pathways of plant differentiation. Differential gene expression includes those genes which are expressed only in certain tissues or particular stages of plant development, as well as those whose expression is altered by environmental factors. The rescue of genomic sequences that direct transgene expression, therefore, is of great interest for studying plant development as it provides a valuable means of analysing the regulatory properties of diverse genomic sequences *in vivo*, and of facilitating isolation of those sequences by virtue of linkage to the foreign DNA. In addition, it may generate cell type-specific developmental markers and may also characterise the genomic sequences that direct specific patterns of gene transcription during development.

REFERENCES

Adam, G. (1984). In : Vectors in virus biology, eds. Mayo, M. & Harrap, K. (Academic, London), Pp. 37-62.

Ahokas, H. (1989). Transfection of germinating barley seed electrophoretically with exogenous DNA. Theor. Appl. Genet. 77 : 469–472.

Ambros, P. ; Matzke, A. and Matzke, M. (1986). Localizaton of *Agrobacterium rhizogenes* T-DNA in plant chromosomes by *in situ* hybridization. EMBO J. 5 : 2073-2077.

Ansorge, W. (1984). Improved system for capillary microinjection into living cells. Exp. Cell Res. 140 : 31-37.

Antonelli, N.M. & Stadler, J. (1989). Chemical methods for efficient direct gene transfer to maize cells : Treatment with polyethylene-glycol or polybrene. J. Genet. & Breed. 43 : 113-122.

Appel, J. ; Fasy, T. ; Kohtz, D. ; Kohtz, J. and Johnson, E. (1988). Asbestos fibres mediate transformation of monkey cells by exogenous plasmid DNA Proc. Natl. Acad. Sci. USA. 85 : 7670–7674.

Armeleo, D. ; Ye, G. ; Jonhston, S. ; Klein, T. ; Shark, K. and Sanford J. (1990). Biolistic nuclear transformation of *Saccharomyces cerevisiae* and other fungi. Curr. Genet. 17 : 97-103.

Athival, R.S. and McBride, O.W. (1977). Serial transfer of a human gene to rodent cells by sequential chromosome-mediated gene transfer. Proc. Natl. Acad. Sci. USA 74 : 2943-2947.

Austin, R. ; Flavell, R. ; Henson , I. and Lowe, H. (1986). Molecular biology and crop improvement. Cambridge Univ. Press. Cambridge.

Avery, O.T. ; Macleod, C.M. and McCarty, M. (1944). Studies on the chemical nature of the substance inducing transformation of pneumococcal types. Induction of transformation by a desoxyribonucleic acid fraction isolated from pneumococcus type III. J. Exp. Med. 79 : 137-158.

Ayad, S.R. and Fox, M. (1969). The implication of repair processes in the mechanism of DNA integration by lymphoma cells. Int. J. Radiat. Biol. 15 : 445-455.

Ballas, N. ; Zakai, N. and Loyter, A. (1987). Transient expression of the plasmid pCaMVCAT in plant protoplasts following transformation with polyethylene-glycol. Experimental cell Res. 170 : 228-234.

Ballas, N. ; Zakai, N. ; Friedberg, D. and Loyter, A. (1988). Linear forms of plasmid DNA are superior to supercoiled structures as active templates for gene expression in plant protoplasts. Plant Mol. Biol. 11 : 517-527.

Behringer, F. and Medford, J. (1992). A plasmid rescue technique for the recovery of plant DNA disrupted by T-DNA insertion. Plant Mol. Biol. Rep. 10 : 190-198.

Benfey, N. and Chan, H. (1989). Regulated genes in transgenic plants. Sci. 244 : 174–188.

Berns, M. ; Aist, J. ; Edwards, J. ; Strahs, K. ; Girton, J. ; McNeill, P.; Rattner, J. ; Kitzes, M. ; Hammer-Wilson, M. ; Liaw, L. ; Siemens, A. ; Koonce, M. ; Peterson, S. ; Brenner, S. ; Burt, J.; walter R.; Bryant, P.; van Dyk, D.; Coulombe, J.; Cahill, T. and Berns, G. (1981). Laster micro-surgery in cell and development biology. Science. 213 : 505-513.

Bevan, M. ; Barnes, W. and Chilton, M.D. (1983). Structure and transcription of the nopaline synthase gene region of T-DNA. Nucleic Acid Res. 11 : 369-385.

Blowers , A. ; Bogorad, L. ; Shark, K. ; Ye, G. and Sanford, J. (1989). Studies on *chlamydomonas* chloroplast transformation : Foreign DNA can be stably maintained in the chromosome. The plant Cell. 1 : 123-132.

Bonham-Smith, P.C. and Bourque, D.P. (1989). Translation of chloroplast—encoded mRNA : Potential initiation and termination signals. Nucleic Acids Res. 17 : 2057-2080.

Boynton, J. ; Gillham, N. ; Harris, E. ; Hosler, J. ; Johnson, A. ; Jones, A. ; Randolph-Anderson, B. ; Robertson, D. ; Klein, T. Shark, K. and Sanford, J. (1988). Chloroplast transformation of *Chlamydomonas* using high velocity microprojectiles. Science. 240 : 1534-1538.

Brisson, N.J. ; Paszkowski, J. ; Penswick, R. ; Gronenborn, B. ; Potrykus, I. and Hohn, T. (1984). Expression of a bacterial gene in plants using a viral vector. Nature. 310 : 511-514.

Bryant, P.E. (1984). Enzymatic restriction of mammalian cell DNA using PvuII and BamHI : Evidence for the double strand break origin of chromosomal aberrations Int. J. Radiat. Biol. 46 : 57-65.

Budar, F., Thia–Toong, L.; Van Montague, M. and Hernalsteens, J. (1986). *Agrobacterium*–mediated gene transfer results mainly in transgenic plants transmitting T–DNA as a Single Mendelian factor. Genet. 114 : 303–313.

Campbell, A. (1960). On the mechanism of the recombinational event in the formation of transducting phages. Virology. 11 : 339-348.

Campbell, A. (1964). Transduction. In The Bacteria, ed. I.C. Gunsalus, R.T. Stanier, 5 : 49-85. New York : Academic.

Cao, J. ; Wang, Y. ; Klein, T. ; Sanford, J. and Wu, R. (1989). Transformation of rice and maize using the biolistic process In Plant Gene Transfer–In Plant Gene Transfer–1989 UCLA Symposium. New York : Alan R. Liss, pp. 21-33.

Carroll, D. (1972). Optical properties of deoxyribonucleic acids-polylysine complexes. Biochemistry 11 : 421-433.

Celis, J.E. (1984). Microinjection of somatic cells with micropipettes : Comparison with other transfer techniques. Biochem. J. 223 : 281-291.

Chasan, R. (1992). Taming the plastid genome. The plan cell 4 : 1-2.

Christou, P. ; McCabe, D.E. and Swain, W.F. (1988). Stable transformation of *soybean* callus by DNA-coated gold particles. Plant Physiology. 87 : 671-674.

Christou, P. and Swain, W.F. (1990). Co-transformation frequencies of foreign genes in *soybean* cell cultures. Theor. Appl. Genet. 79 : 337-341.

Clark, A.J. ; Cowper, A. ; Wallace, R. ; Wright, G. and Simons, J.P. (1992). Rescuing transgene expression by co-integration. Bio/Technology 10 : 1450-1454.

Coen, E. ; Romero, J. ; Doyle, S. ; Elliott, R. ; Murphy, G. and Carpenter, R. (1990). Floricaula : A homeotic gene required for flower development in *Antirrhinum majus.* Cell. 63 : 1311-1322.

Cooley, L. ; Kelley, R. and Sprading, A. (1988). Insertional mutagenesis of the *Drosophila* genome with single P. elements. Science. 239 : 1121-1128.

Clark, A.J. (1973). Recombination deficient mutant of *E. coli* and other bacteria. Ann. Rev. Genet. 7 : 67-86.

Czernilofsky, A.P. ; Hain, R. ; Baker, B. and Writz, U (1986a). Studies of the structure and functional organization of foreign DNA integrated into the genome of *Nicotiana tabacum.* DNA 5 : 473-482.

Czernilofsky, A.P. ; Herrera-Estrella, L. ; Lorz, H. ; Goyvaerts, E. ; Baker, B.J. and Schell, J. (1986b). Fate of selectable marker DNA integrated into the genome of *Nicotiana tabacum.* DNA 5 : 101-113.

Danniell, H. ; Vivekananda, J. ; Nielsen, B.L. ; Ye, G.N. ; Tewari, K.K. and Sanford, J.C. (1990). Transient foreign gene expression in chloroplasts of cultured tobacco cells after biolistic delivery of chloroplasts. Proc. Natl. Acad. Sci. USA 87 : 88-92.

Danniell, H. ; Krishnan, M. and McFadden, B.F. (1991). Transient expression of β-glucuronidase in different cellular compartments following biolistic delivery of foreign DNA into wheat leaves and calli. Plant Cell Rep. 9 : 615-619.

Davey, M.R. ; Cocking, E.C. ; Freeman, J. ; Pearce, N. and Tudor, I. (1980). Transformation of *Petunia* protoplasts by isolated Agrobacterium plasmids. Plant Sci. Lett. 18 : 307-313.

Dean, C.; Jones, J.; Favrean, M.; Dunamuir, P. and Bedbrook, J. (1988). Influence of flanking sequences on variability in expression levels of an introduced gene in transgenic tobacco plants. Nucl. Acids, Res. 16 : 9267–9283.

De Block, M. ; Schell, J. and Van Montagu, M. (1985). Chloroplast transformation by *Agrobacterium tumefaciens.* EMBO J. 4 : 1367-1372.

Degnen, G.E. ; Miller, I.L. ; Adelberg, E.A. and Eisenstadt, J.M. (1976). Chromosome-mediated gene transfer between closely related strains of cultured mouse cells. Proc. Natl. Acad. Sci. USA 73 : 2838-2842.

Degnen, G.E. ; Miller, I.L. ; Adelberg, E.A. and Eisenstadt, J.M. (1977). Overexpression of an unstably inherited gene in cultured mouse cells. Proc. Natl. Acad. Sci. USA 74 : 3956-3959.

Dekeyser, R. ; Claes, B. ; Rycke, R. ; Habets, M. and van Montagu, M. (1990). Transient gene expression in intact and organized rice tissues. The plant Cell. 2 : 591-602.

De la Pena, A. ; Lorz, H. and Schell, J. (1987). Transgenic rye plants obtained by injecting DNA into young floral tillers. Nature. 325 : 274-276.

Delauney, A. ; Tabaeizaden, Z. and Verma, D. (1988). A stable bifunctional antisense transcript inhibiting gene expression in transgenic plants. Natl. Acad. Sci. USA 85 : 4300-4304.

De Lozanne, A. and Spudich, J.A. (1987). Disruption of the *Dictyosteliunmyosin* heavy chain gene by homologous recombination. Science 236 : 1086–1091.

Dennis, E. and Brettell, R. (1990). DNA methylation of maize transposable elements is correlated with activity. Phil. Trans. R. Soc. Lond. B. 326 : 217–229.

Depicker, A. ; Stachel, S. ; Dhaese, P. ; Zambryski, P. and Goodman, H. (1982). Nopaline synthase : transcript mapping and DNA sequence. J. Mol. Appl. Genet. 1 : 561-573.

Deroles, S. and Gardner, R. (1988). Analysis of the T–DNA structure in a large number of transgenic petunias generated by *Agrobacterium* mediated transformation. Plant Mol. Biol. 11 : 365–377.

D'Halluin, K. ; Bonne, E. ; Bossut, M. ; Beuckeleer, M.D. & Leemans, J. (1992). Transgenic maize plans by tissue electroporation. The plant cell. 4 : 1495-1505.

Dove, W.F. (1968). The genetics of the lambdoid phages. Ann. Rev. Genet. 2 : 305-340.

Draper, J. ; Davey, M. and Freeman, J. (1986). Delivery systems for transformation of plant protoplasts. In Plant Cell Culture Technology, (Yeoman, M. ed) Pp. 271-301 Blackwell Scientific Publications, Oxford.

Dubbs, D.R. ; Rachmeler, M. and Kit, S. (1974). Recombination between temperature-sensitive mutants of simian virus 40. Virology 57 : 161-174.

Duin, M.V. ; Westerveld, A. and Hoeijmarkers, J.H. (1985). U.V. stimulation of DNA-mediated transformation of human cells. Mol. & Cell Biol. 5 : 734-741.

Elmer, J. ; Browing, C. ; Bisaro, D. and Rogers, S. (1988). *Agrobacterium*-mediated inoculation of plants with tomato golden mosaic virus DNAs. Piant Mol. Biol. 10 : 225-234.

Falkow, S. (1975). Infection multiple drug resistance. London : Pion.

Feldmann, K. ; Christianson, M. and Quatrano, R. (1989). A dwarf mutant of *Arabidopsis* generated by T-DNA insertion mutagenesis. Science. 243 : 1351-1354.

Fitch, M.; Manshardt, R. ; Gonsalves, D. ; Slightom, J. and Sanford, J. (1990). Stable transformation of papaya via microprojectile bombardment. Plant Cell Rep. 9 : 189-194.

Fraley, R. ; Dellaporta, S. and Papahadjopoulos, D. (1982). Liposome-mediated delivery of tobacco mosaic virus RNA into *Petunia* protoplasts. A sensitive assay for monitoring liposome-protoplast interactions. Proc. Natl. Acad. Sci. USA. 79 : 1859-1863.

Fraley, R. (1983). Liposome-mediated delivery of tobacco masaic virus RNA into tobacco protoplasts. Plant Mol. Biol. 2 : 5-14.

Franklin, N.C. (1971). Illegitimate recombination. In the bacteriophage lambda, ed., A.D. Hershey, Pp. 175-194. Cold Spring Harbor, NY : Cold Spring Harbor Lab.

Fromm, M. ; Taylor, L. and Walbot, V. (1985). Expression of genes transferred into monocot and dicot plant cells by electroporation. Proc. Natl. Acad. Sci. USA 82 : 5824-5828.

Fromm, M. ; Taylor, L. and Walbot, V. (1986). Stable transformation of maize after gene transfer by electroporation. Nature. 319: 791-793.

Fromm, M. ; Morrish, F. ; Armstrong, C. ; Williams, R. ; Thomas, J. and Klein, T. (1990). Inheritance and expression of chimeric genes in the progeny of transgenic maize plants. Biotechnology. 8 : 833–839.

Gad, A. ; Rosenberg, N. and Altman, A. (1990). Liposome-mediated gene delivery into plant cells. Physiol. Plant. 79 : 177-183.

Gallois, P. ; Lindesy, K. ; Malone, R. ; Kreis, M. and Jones, M. (1992). Gene rescue in plants by direct gene transfer of total genomic DNA into protoplasts. Nucleic Acid Res. 20 : 3977-3982.

Gharti-Chhetri, G.B. ; Cherdshewasart, W. ; Dewulf, J. ; Paszkowski, J. ; Jacobs, M. & Negrutiu, I. 1990. Hybrid genes in the analysis of transformation conditions. 3. Temporal / spatial fate of NPT II gene integration, its inheritance and factors affecting these process in *Nicotiana plumbaginfolia.* Plant Mol. Biol. 14 : 687-696.

Gierl, A. and Saedler, H. (1992). Plant-transposable elements and gene tagging. Plant Mol. Biol. 19 : 39-49.

Gilley, D. ; Preer, J.R. ; Aufderheide, K.J. and Polisky, B. (1988). Autonomous replication and addition of telomerelike sequences to DNA microinjected into *Paramecium tetraurelia* macronuclei. Mol. Cell. Biol. 8 : 4765-4772.

Gingery, R. and Echols, H. (1967). Mutants of bacteriophage unable to integrate into the host chromosome. Proc. Natl. Acad. Sci. USA 58 : 1507-1514.

Gleba, Y. and Sytnik, K. (1984). "Protoplast Fusion-Genetic Engineering in Higher Plants". Springer Verlag, Berlin.

Golds, T. ; Maliga, P. and Koop, H-U. 1993. Stable plastid transformation in PEG-treated protoplasts of *Nicotiana tabacum.* Bio/Technology 11 : 95-97.

Goodwin, P.B. and Erwee, M.G. (1985). Intranuclear transport studied by microinjection method. In Botanical Microscopy. (A.W. Robards, ed.), pp. 335-358. Oxford Univ. Press.

Gordon-Kamm, W. ; Spencer, T. ; Mangano, M. ; Adams, T. ; Daines, R. ; Start, W. ; O'Brien, J. ; Chamber, S. ; Adams, J. ; Willetts, N. ; Rice, T. ; Mackey, C. ; Krueger, R. ; Kausch, A. and Lemaux, P. (1990). Transformation of maize cells and regeneration of fertile transgenic plants. The Plant Cell 2 : 603-618.

Gorman, C.M. ; Moffatt, L.F. and Howard, B.H. (1982). Recombinant genomes which express chloramphenicol acetyl transferase in mammalian cells. Mol. Cell. Biol. 2 : 1044-1051.

Gottesman, M.E. and Yarmolinsky, M.B. (1968). Integration-negative mutants of bacteriophage lambdo. J. Mol. Biol, 31 : 487-505.

Greulich, K. ; and Weber, G. (1992). The light microscope its way from analytical to a preparative tool. Journal of microscopy. 167 : 127-151.

Griesbach, R.J. (1983). Protoplast microinjection. Plant Mol. Biol. Rep. 1 : 32-37.

Griesbach, R.J. (1987). Chromosome-mediated transformation via microinjection. Plant Sci. 50 : 69-77.

Griffith, F. (1928). Significance of pneunmococcal types. J. Hyg., Camb. 27 : 113-159.

Grimsley, N. ; Hohn, B. ; Hohn, T. and Walden, R.(1986). Agroinfection, an alternative route for viral infection of plants using the Ti plasmid. Proc. Natl. Acad. Sci. USA 83 : 3282-3286.

Grimsley, N. ; Hohn, B. ; Hohn, T. and Walden, R. (1986b). Recombination in a plant virus : template-switching in cauliflower mosaic virus. EMBO J. 5 : 641-646.

Grimsley, N. and Bisaro, D. (1987). Agroinfection. In Schell, J. ; Hohn, T. (eds). Plant DNA Infectious agents. Plant Gene Research. Springer, New York and Vienna, Pp. 87-107.

Guang-Ning, Ye. ; Danniell, H. and Sanford, J.C. (1990). Optimization of delivery of foreign DNA into higher-plant chloroplasts. Plant Mol. Biol. 15 : 809-819.

Hain, R. ; Stabel, P. ; Czernilofsky, A.P. ; Steinbiss, H.H. ; Herrera-Estrella, L. and Schell, J. (1985). Uptake, integration, expression and genetic transmission of a selectable chimeric gene by plant protoplasts. Mol. Gen. Genet. 199 : 161-168.

Harrison, B. and Murant, A. (1984). In Vectors in virus biology, eds. Mayo, M. & Harrap, K. (Academic, London). Pp. 1-36.

Hall, J.D. Featherston, J.D. and Almy, R.E. (1980). Evidence for repair of ultraviolet light-damaged *Herpes* virus in human fibroblasts by recombination mechanism. Virology. 105 : 490-500.

Halliwell, R.S. and Gazaway, W.S. (1975). Quality of microinjection tobacco mosaic virus required for infection of single cultured tobacco cells. Virology. 65 : 583-587.

Hammer, R.E. ; Pursel, V.G. ; Rexnroad, C.E. ; Wall, R.J., Bolt, D.J.; Ebert, K.M. ; Palmiter, R.D. and Brinster, R.L. (1985). Production of transgenic rabbits, sheep and pigs by microinjection. Nature. 315 : 680-683.

Harrison, B. ; Mayo, M. and Baulcombe, D. (1987). Virus resistance in transgenic plants that express cucumber mosaic virus satellite RNA. Nature. 328 : 799-802.

Hasnain, S. ; Ganesan, K. and Upadhyaya, K. (1980). Mechanism of enhancement of DNA-uptake by polycations : Effect of polycations on DNA, DNAs & plasma membrane. Indian J. of Experimental Biology. 18 : 1230-1232.

Hayes, R. ; Petty, I. ; Coutts, R. and Buck, K. (1988b). Gene amplification and expression in plants by a replicating geminivirus vector. Nature. 334 : 179-182.

Hayles, J. and Nurse, P. (1992). Genetics of the fission yeast *Schizosaccharomyces pombe* Ann. Rev. Genet. 26 : 273-402.

Heintz, N. ; Sive, H. and Roeder, R. (1983). Regulation of human histone gene expression : Kinetics of accumulation and changes in the rate of synthesis and in the half-lives of individual histone mRNAs during the hela cell cycle. Mol. Cell. Biol. 3 : 539-550.

Heitman, J. and Model, P. (1987). Site-specific methylases induce the SOS DNA repair response in *E. coli.* J. Bacteriology. 169 : 3243-3250.

Helmer, G. ; Gasadaban, M. ; Bevan, M.W. ; Kayes, L. and Chilton, M.D. (1984). A new chimeric gene as a marker for plant transformation : the expression of *E. coli* β-galactosidase in sunflower and tobacco cells. Bio/Technology. 2 : 520-527.

Herrera-Estrella, L. ; Depicker, A. ; Van Montagu, M. and Schell, J. (1983). Chimeric genes are transferred and expressed in plants using a Ti plasmid-drived vector. Nature. 303 : 209-213.

Heyer, W.D. ; Sipiczki, M. and Kohli, J. (1986). Replicating plasmids in *Schizosaccharomyces pombe* : improvement of symmetric segregation by a new genetic element. J. Mol. Biol. 6 : 80-89.

Hinnen, A. ; Hicks, J. and Fink, G. (1978). Transformation of yeast. Proc. Natl. Acad. Sci. USA. 75 : 1929-1933.

Hoch, B. ; Maier, R.M. ; Appel, K. ; Igloi, G.L. and Kossel, H. (1991). Editing of a chloroplast mRNA by creation of an initiation codon. Nature. 353 : 178-180.

Hoekema, A. ; Hooykaas, P. and Schilperoort, R. (1984). Transfer of octopine T-DNA segment to plant cells mediated by different types of *Agrobacterium* tumor or root inducing plasmids : generality of virulences systems. J. Bact. 158 : 383-385.

Holliday, R. (1987). The inheritance of epigenetic defects. Sci. 23 : 163–169.

Hope, I. (1991). "Promoter Trapping" in *Caenorhabditis elegans*. Development 113 : 399–408.

Hotchkiss, R.D. (1974). Models of genetic recombination. Ann. Rev. Genet. 8 : 445-468.

Hu, C.–Y. ; Chee, P.P. ; Chesney, R.H. ; Zhon, J.H. ; Miller, P.D. and O'Brien, W.T. (1990). Intrinsic GUS-like activity. In : Nijkamp. H.J. ; Van der plas, L.H. and Van Aartrijk, J. (eds.) progress in plant cellular and molecular biology. Kluwer, Dordrecht, pp. 195-200.

Hughes, K. ; Case, M.E. ; Geever, R. ; Vapnek, D. and Giles, N.H. (1983). Chimeric plasmid that replicates autonomously in both *E. coli* and *Neurospora crassa*. Proc. Natl. Acad. Sci. USA. 80 : 1053-1057.

Jefferson, R.A. ; Burgess, S.M. and Hirsh, D. (1986). β-glucuronidase from *E. coli* as a gene-fusion marker. Proc. Natl. Acad. Sci. USA. 83 : 8447-8451.

Jefferson, R.A. ; Kavanagh, T.A. & Bevan, M.W. (1987). GUS fusions: β-glucuronidase as a sensitive and versatile gene fusion marker in higher plants. EMBO J. 6 : 3901-3907.

Joersbo, M. (1990). A method for introducing molecules, particularly genetic material, into plant cells. Danish patent application 3251/89.

Joersbo, M. and Brunstedt, J. (1990a). Direct gene transfer to plant protoplasts by mild sonication. Plant Cell Reports. 9 : 207-210.

Joersbo, M. and Brunstedt, J. (1990b). Direct gene transfer to plant protoplasts by alternating, rectangular and exponentially decaying pulses. Plant Cell Reports. 8 : 701-705.

Joersbo, M. and Brunstedt, J. (1992). Sonication : A new method for gene transfer to plants. Physiol. Plant. 85 : 230-234.

Johnston, S. ; Anziano, P. ; Shark, K. ; Sanford, J. and Butow, R. (1988). Transformation of yeast mitochondria by bombardment of cells with microprojectiles. Science. 240 : 1538-1541.

Jones, J.; Gilbert, D.; Grady, K. and Jorgensen, R. (1987). T–DNA structure and gene expression in petunia plants transformed by *Agrobacterium tumefaciens* 58 derivatives. Mol. Gen. Genet. 207 : 478–485.

Jones, J.D. ; Dunsmuir, P. and Bedbrook, J. (1985). High level expression of introduced chimeric genes in regenerated transformed plants. EMBO J. 4 : 2411-2418.

Jongsma, M. ; Korrnneef, M. ; Zabel, P. and Hille, J. (1987). Tomato protoplast DNA transformation : Physical linkage and recombination of exogenous DNA sequences. Plant Mol. Biol. 8 : 383-394.

Jorgensen, R. (1990). Altered gene expression in plants due to *trans* interactions between homologous genes. Trends in Biotechnol. 8 : 340–344.

Junker, B. (1987). Studien zur transformation und *in vitro* Kultur von Mais (*Zea mays L.*), Ph.D. Thesis, Univ. Koln, FRG.

Junker, B. ; Zimny, J. ; Luhrs, R. and Lorz, H. (1987). Transient expression of chimeric genes in dividing and non-dividing cereal protoplasts after PEG-induced DNA uptake. Plant Cell Reports. 6 : 329-332.

Kaeppler, H. ; Gu, W. ; Somers, D. ; Rines, H. and Cockburn, A. (1990). Silicon carbide fiber-mediated DNA delivery into plant cells. Plant Cell Reports 8 : 415-418.

Kaeppler, H.; Somers, D.; Rines, H. and Cockburn, A. (1992). Silicon carbide fiber-mediated stable transformation of plant cells. Theo. Appl. Genet. 84 : 560–566.

Kapler, G.M. ; Coburn, C.M. and Beverley, S.M. (1990). Stable transfection of the human parasite *Leishmania major* delineates a

30-kilobase region sufficient for extrachromosomal replication and expression. J. Cell. Biol. 10 : 1084-1094.

Kertbundit, S.; De Greve, H.; Deboeck, F. and van Montague, M. (1991) *In vivo* random β–glucuronidase gene fusions in *Arabidopsis thaliana.* Proc. Natl. Acad. Sci. USA 88 : 5212–5216.

Klee, H. ; Hayford, M. and Rogers, S. (1987). Gene rescue in plants: A model system for "shotgun" cloning by re-transformation. Mol. Gen. Genet. 210 : 282-287.

Klein, T ; Wolf, E. ; Wu, R. and Sanford, J. (1987). High-velocity microprojectiles for delivering nucleic acids into living cells. Nature. 327 : 70-73.

Klein, T. ; Fromm, M. ; Weissinger, A. ; Tomes, D. ; Schaaf, S. ; Sleeten, M. and Sanford, J. (1988a). Transfer of foreign genes into intact maize cells using high velocity microprojectiles. Proc. Natl. Acad. Sci. USA. 85 : 4305-4309.

Klein, T.M. ; Gradziel, T. ; Fromn, M.E. and Sanford, J.C. (1988b). Factors influencing gene delivery into *Zea mays* cells by high-velocity microprojectiles. Bio/Technology. 6 : 559-563.

Klein, T. ; Harper, E. ; Svab, Z. ; Sanford, J. ; Fromm, M. and Maliga, P. (1988c). Stable genetic transformation of intact *Nicotiana* cells by the particle bombardment process. Proc. Natl. Acad. Sci. USA. 85 : 8502-8505.

Klein, T.M. ; Kornstein, L. ; Sanford, J.C. and Fromm, M.E. (1989). Genetic transformation of maize cells by particle bombardment. Plant Physiol. 91 : 440-444.

Klein, T. ; Goff, S. ; Roth, B. and Fromm, M. (1990). Applications of the particle gun in plant biology. In the proceedings of the VII international congress on plant tissue and cell culture. Amsterdam, The Netherlands, June 24-29.

Klein, T ; Roth, B. and Fromm, M. (1991). Regulation of anthocyanin biosynthetic genes introduced into intact maize tissues by microprojectiles. Proc. Natl. Acad. Sci. USA. 86 : 6681-6685.

Klobutcher, L.A. and Ruddle, F.H (1979). Phenotype stabilisation and integration of transferred material in chromosome-mediated gene transfer. Nature. 280 : 657-660.

Klobutcher, L.A. ; Miller, C.L. and Ruddle, F.H. (1980). Chromosome-mediated gene transfer results in two classes of unstable transformants. Proc. Natl. Acad. Sci. USA. 77 : 3610-3614.

Kloti, A. ; Iglesias, V. ; Wunn, J. ; Burkhardt, P. ; Datta, S. and Potrykus, I. (1993). Gene transfer by electroporation into intact scutellum cells of wheat embryos. Plant Cell Reports. 12 : 671-675.

Knight, C. ; Sawahel, W. and Cove, D. (1993). Gene flux in the moss *Physcomitrella patens*–the fate of foreign DNA in three classes of transformants. J. Experimental Botany. 44 : 32.

Kohler, F. ; Golz, C. ; Eapen, S. and Schieder, O. (1987). Influence of plant cultivar and plasmid DNA on transformation rates in tobacco and moth bean. Plant Science. 53 : 87-91.

Kohler, F. ; Afsar, K. ; Golz, C. ; Eapen, S. ; Kohn, H. and Schieder, O. (1987). Stable transformation of moth bean *Vigna aconitifolia* via direct gene transfer. Plant Cell Rep. 6 : 313-317.

Kohler, F. ; Cardon, G. ; Pohlman, M. ; Gill, R. and Schieder, O. (1989). Enhancement of transformation rates in higher plants by low-dose irradiation : are DNA repair systems involved in the incorporation of exogenous DNA into the plant genome ? Plant Mol. Biol. 12 : 189-199.

Kohler, F. ; Benediktsson, I ; Cardon, G. ; Andreo, C. and Schiedor, O. (1990). Effect of various irradiation treatments of plant protoplasts on the transformation rates after direct gene transfer. Theor. Appl. Genet. 79 : 679-685.

Knocz, C. and Shell, J. (1986). The promoter of TL-DNA gene 5 controls the tissue-specific expression of chimaeric genes carried by a novel type of *Agrobacterium* binary vector. Mol. Gen. Genet. 204 : 383-396.

Koncz, C. ; Martini, N. ; Mayerhofer, R. ; Koncz-Kalman, Z. ; Korber, H. ; Redei, G. and Schell, J. (1989). High-frequency T-DNA-mediated gene tagging in platns. Proc. Natl. Acad. Sci. USA. 86 : 8647-8651.

Kornberg, A. (ed) (1974). DNA synthesis. San Francisco : Freeman.

Krens, F.A. ; Molendijk, L. ; Wullems, G.J. and Schilperoort, R.A. (1982). In vitro transformation of plant protoplasts with Ti-plasmid DNA. Nature. 296 : 72-74.

Krol, A.; Mur, L.; Beld, M. ; Mol, J. and Stuitje, A. (1990). Flavonoid genes in *Petunia* addition of a limited number of gene copies may lead to a suppression of gene expression. Plant Cell 2 : 291–299.

Kuspa, A. and Loomis, W. (1992). Tagging developmental genes in *Dictyostelium* by restriction enzyme-mediated integration of plasmid DNA. Proc. Natl. Acad. Sci. USA. 89 : 8803-8807.

Lack, S.A. (1978). Mechanisms of DNA uptake by cells. In : Genetic interaction and gene transfer. Brookhaven symp. Biol. 29.

Lawrence, W.A. and Davies, R.D. (1985). A simple method for the microinjection and culture of protoplasts at very low densities. Plant Cell Rpts. 4 : 33-35.

Lee, K.Y. ; Lund, P. ; Lowe, K. and Dunsmuir, P. (1990). Homologous recombination in plant cells after *Agrobacterium*–mediated transformation. The Plant Cell. 2 : 415-425.

Lindsey, K.; Wei, W.; Clarke, M.; McArdle, H.; Rooke, L. and Topping, J. (1993). Tagging genomic sequences that direct transgene expression by activation of a promoter trap in plants. Transgenic Res. 2 : 33–47.

Lindsey, K. and Jones, M. (1990) Electroporation of cells. Physiol. Plant. 79 : 168-172.

Linn, F.; Heidmann, I.; Saedler, H. and Meyer, P. (1990). Epigenetic changes in the expression of the maize A1 gene in petunia hybrida: Role of numbers of integrated gene copies and state of methylation. Mol. *Gen. Genet.* 222 : 329–336.

Lonsdale, D. ; Onde, S. and Cuming, A. (1990). Transient expression of exogenous DNA in intact, viable wheat embryos following particle bombardment. Journal of Experimental Botany. 41 : 1161-1165.

Lorz, H. ; Baker, B. and Schell, J. (1985). Gene transfer to cereal cells mediated by protoplast transformation. Mol. Gen. Genet. 199 : 179-182.

Lucas, W. ; Lansing, A. ; De Wet, J. and Walbot, V. (1990). Introduction of foreign DNA into walled plant cells via liposomes injected into the vacuole : A preliminary study. Physiol. Plant. 79 : 184-189.

Madore, M.A. ; Oross, J.W. and Lucas, W.J. (1986). Symplasmic transport in *Ipomea tricolor* source leaves : demonstration of functional symplastic connections from mesophyll minor veins by a novel dye-tracer method. Plant Physiol. 82 : 432-442.

Mandal, A. ; Lang, V. ; Orezyk, W. and Palva, E. (1993). Improved efficiency for T-DNA-mediated transformation and plasmid rescue in *Arabidopsis thaliana.* Theor. Appl. Genet. 86 : 621-628.

Maliga, P. (1993). Towards plastid transformation in flowering plants. Trends in Biotechnology. 11 : 101-107.

Marini, N.J. ; Etkin, L.D. and Benbow, R.M. (1988). Perisistence and replication of plasmid DNA microinjected into early embryos of Xenopus laevis. Dev. Biol. 127 : 421-434.

Mathias, R. (1987). In "Genetic Engineering-Principles and Methods" (J.K. Setlow, ed.), Vol. 9, pp. 199-227. Plenum Prss, New York.

Matzke, M.; Primig, M.; Trnovsky, J. and Matzke, A. (1989). Reversible methylation and inactivation of marker genes in sequentially transformed tobacco plants. EMBO J. 8 : 643–649.

Matzke, M. and Matzke, A. (1990). Gene interactions and epigenetic variations in transgenic plants. Dev. Genet. 11 : 214–223.

Meyerhofer, R. ; Koncz-Kalman, Z. ; Nawrath, C. ; Bakkeren, G. ; Crameri, A. ; Angelis, K. ; Redei, G. ; Schell, J. ; Hohn, B. and Koncz, C. (1991). T-DNA integration : a mode of illegitimate recombintion in plants. EMBO J. 10 : 697-704.

McBride, O.W. and Ozer, H.L. (1973). Transfer of genetic information by purified metaphase chromosomes. Proc. Natl. Acad. Sci. USA. 70 : 1258-1262.

McCabe, D. ; Swain, W. ; Martinell, B. and Christou, P. (1988). Stable transformation of soybean (*Glycine max*) by particle acceleration. Biotechnology. 6 : 923-926.

McMahon, A.P. ; Flytzanis, C.N. ; Hough-Evans, B.R. ; Katula, K.S.; Britten, R.J. and Davidson, E.H. (1985). Introduction of cloned DNA into sea urchin egg cytoplasm : replication and presistence during embryogenesis. Dev. Biol. 108 : 420-430.

Meilhoc, E. ; Masson, J.M. and Teissie, J. (1990). High efficiency transformation of intact yeast cells by elecric field pulses. Bio/Technology. 8 : 223-227.

Mello, C.C. ; Kramer, J.M. ; Stinchcomb, D. and Ambros, V. (1991). Efficient gene transfer in *C. elegans* : extrachromosomal maintenance and integration of transforming sequences. The EMBO J. 10 : 3959-3970.

Mendel, R. ; Muller, B. ; Schulze, J. ; Kolesnikov, V. and Zelenin, A. (1989). Delivery of foreign genes to intact barley cells by high-velocity microprojectiles. Theor. Appl. Genet. 78 : 31-34.

Meyer, P. ; Walgenbach, E. ; Bussmann, K. ; Homb reacher, G. and Saedler, H. (1985). Synchronized tobacco protoplasts are efficiently transformed by DNA. Mol. Gen. Genet. 201 : 513-518.

Meyer, P. ; Kartzke, S. ; Niedenhof, I. ; Heidmann, I.; Bussmann, K. and Saedler, H. (1988). A genomic DNA segment from *Petunia hybrida* leads to increase transformation frequenceis and simple integration patterns. Proc. Natl. Acad. Sci. USA. 85 : 8568-8572.

Miller, C.K. and Temin, H.M. (1983). High-efficiency ligation and recombination of DNA fragments by vertebrate cells. Science. 220 : 606-609.

Morgan, W.F. and Cleaver, J.E. (1983). Effect of 3-aminobenzamide on the rate of ligation during the repair of alkylated DNA in human fibroblasts. Cancer Res. 43 : 3104-3107.

Morikawa, H. and Yamada, Y. (1985). Capillary microinjection into protoplasts and interanuclear localization of injected materials. Plant Cell Physiol. 26 : 229-236.

Morikawa, H ; Lida, A. ; Matusi, C. ; Ikegami, M. and Yamada, Y. (1986). Gene transfer into intact plant cells by electroinjection through cell walls and membranes. Gene. 41 : 121-124.

Morikawa, H. ; Lida, A. and Yamada, Y. (1989). Transient expression of foreign genes in plant cells and tissues obtained by a simple biolistic device. Appl. Microbial Biotechnol. 31: 320-322.

Muller, A.; Mendel, R.; Schiemann, J.; Simones, C. and Inze, D. (1987). High meiotic stability of a foreign gene introduced into tobacco by *Agrobacterium* mediated transformation. Mol. Gen. Genet. 207 : 171–175.

Mullet, J.E. (1988). Chloroplast development and gene expression. Annu. Rev. Plant. Physiol. Plant Mol. Biol. 39 : 475-502.

Mullineauz, P. ; Donson, J. ; Morris-Krsinich, B. ; Boulton, M. and Davies, J. (1984). The nucleotide sequence of maize streak virus DNA. EMBO J. 3 : 3063-3068.

Mullineaux, Boulton, M. ; Bowyer, P. ; Van der Vlugt, R. ; Marks, M.; Donson, J. and Davies, J. (1988). Detection of a non-structural protein of MW 11000 encoded by the virion DNA of maize streak virus. Plant Mol. Biol. 11 : 57-66.

Nagy, F.; Morelli, G.; Fraley, R.; Rogers, S. and Chua, N. (1985). Photoregulated expression of a pea *rbc* S gene in leaves of transgenic plants. EMBO J. 4 : 3063–3068.

Napoli, C.; Lemieux, C. and Jorgensen, R. (1990). Introduction of a Chimeric *Chalcone Synthase* gene into *petunia* results in reversible co-suppression of homologous gene in *trans.* Plant Cell 2 : 279–289.

Negrutiu, I. ; De Wrouwer, D. ; Driks, R. and Jacobs, M. (1985). Amino acid auxotrophs from protoplasts cultures of *Nicotiana plumbaginifolia* viviani. 1. BuDR enrichment selection, plant regeneration and general characterization. Mol Gen. Genet. 199 : 330-337.

Negrutiu, I. ; Shillito, R. ; Potrykus, I. ; Biasini, G. and Sala, F. (1987). Hybrid genes in the analysis of transformation conditions. I. Setting up a simple method for direct gene transfer in plant protoplasts. Plant Mol. Biol. 8 : 363-373.

Negrutiu, I. ; Dewulf, J. ; Pietrzark, M. ; Bottermann, J. ; Rietveld, E.; Wurzer-Figurelli, E.M. ; Ye, D. & Jacobs, M. (1990). Hybrid gene in the analysis of transformation conditions. II

Transient expression vs stable transformation–analysis of parameters influencing gene expression levels and transformation efficiency. Physiol. Plant. 79 : 197-205.

Neumann, E. ; Schaefer-Ridder, M. ; Wang, Y. and Hofschneider, P.H. (1982). Gene transfer mouse lyoma cells by electroporation in high electric fields. The EMBO Journal. 1 : 841-845.

Newlon, C.S. (1988). Yeast chromosome replication and segregation . Microb. Rev. 52 : 568-601.

Nims, R.C. ; Halliwell, R.S. and Rosberg, D.W. (1967a). Wound healing in cultured tobacco cells following microinjection. Protoplasma. 64 : 305-314.

Nims, R.C. ; Halliwell, R.S. and Rosberg, D.W. (1967b). Disease development in cultured cells of *Nicotiana tabacum* L. var. samsun NN injected with tobacco mosaic virus. Cytologia. 32 : 224-235.

Nishiguchi,M. ; Sato, T. and Motoyoshi, F. (1987). An improved method for electroporation in plant protoplasts : Infection of tobacco protoplasts by tobacco mosaic virus particles. Plant Cell Reports. 6 : 90-93.

Noyer-Weidner, M. ; Diaz, R. and Reiners, L. (1986). Cytosine-specific DNA modification interferes with plasmid establishment in *E. coli* K12 : involvement of *rg/B*. Mol. Gen. Genet. 205 : 469-475.

Oard, J. ; Paige, D. ; Simmonds, J. and Gradizel, T. (1990). Transient gene expression in maize, rice and wheat cells using an airgun apparatus. Plant. Physiol. 92 : 334-339.

Odell, J.; Nagy, F. and Chua, N. (1987). Variability in 35S promoter expression between independent transformants. In "Plant Gene Systems and their Biology" (eds : Key, J. and McIntosh, L.) Vol. 62, Alan R. Liss, New York.

Okada, K. ; Takebe, I. and Nagata, T. (1989). Expression and integration of genes introduced into highly synchronized plant protoplasts. Mol. Gen. Genet. 205 : 398-403.

Ott, R. and Chua, N. (1990). Enhancer sequences from *Arabidopsis thaliana* obtained by library transformation of *Nicotiana tabacum*. Mol. Gen. Genet. 223 : 169-179.

Ow, D.W. ; Wood, K.V. ; Deluca, M. ; Dwewet, J.R. ; Helinski, D.R. and Howell, S.H. (1986). Transient and stable expression of the firefly luciferase gene in plant cells and transgenic plants. Science. 234 : 856-859.

Paszkowski, J. ; Shillito, R. ; Saul, M. ; Mandak, V. ; Hohn, T. ; Hohn, B. and Potrykus, I. (1984). Direct gene transfer to plants. EMBO J. 3 : 2717-2722.

Paszkowski, J. ; Baur, M. ; Bogucki, A. and Potrykus, I. (1988). Gene targeting in plants. The EMBO journal. 7 : 4021-4026.

Paszkowski, J. ; Saul, M. and Potrykus, I. (1989). Plant gene vectors and genetic transformation : DNA-mediated direct gene transfer to plants. In Cell Culture and Somatic Cell Genetics of Plants. (J. Schell & I. Vasil, eds) Vol 6, pp. 52-68.

Peerbolte, R. ; Krens, F.A. ; Mans, R.M. ; Floor, M. ; Hoge, J.H. ; Wullems, G.J. and Schilperoort, R.A. (1985). Transformation of plant protoplasts with DNA : Co-transformation of non-selected calf thymus DNA sequences. Plant Mol. Biol. 5 : 235-246.

Perucho, M; Hanahan, D. and Wigler, M. (1980). Genetic and physical linkage of exogenous sequences in transformed cells. Cell. 22 : 309-317.

Perucho, M.; Hanahan, D. ; Lipsich, L. and Wigler, M. (1980). Isolation of chicken thymidine kinase gene by plasmid rescue. Nature. 285 : 207–210.

Petit, A. ; David, C. ; Dahl, G. ; Ellis, J. ; Guyon, P. ; CasseDelbart, F. and Tempe, J. (1983). Further extension of the opine concepts : plasmids in *Agrobacterium rhizogenes* cooperate for opin degradation. Mol. Gen. Genet. 190 : 204-214.

Pietrzak, M. ; Shillito, R.D. : Hohn T. and Potrykus, I. (1986). Expression in plants of two bacterial antibiotic resistance genes after protoplast transformation with a new plant expression vector. Nucleic Acids Res. 14 : 5857-5868.

Postel, E.H. (1985). Enhancement of genetic transformation frequencies of mammalian cell cultures by damage to the cell DNA. Mol. Gen. Genet. 210 : 136-139.

Potrykus, I. ; Shillito, D. ; Saul, M.W. and Paszkowski, J. (1985a). Direct gene transfer. State of the art and future potential. Plant Molecular Biology Reporter. 3 : 117-128.

Potrykus, I ; Paszkowski, J. ; Saul, M. ; Petruska, I. and Shillito, R. (1985b). Molecular and general genetics of a hybrid foreign introduced into tobacco by direct gene transfer. Mol. Gen. Genet. 199 : 169-177.

Potrykus, I. ; Saul M. ; Petruska, I. ; Paszkowski, J. and Shillito, R. (1985c). Direct gene transfer to cells of a graminaceous monocot. Mol. Gen. Genet. 199 : 183-188.

Potrykus, I. (1990). Gene transfer to plants : assessment and perspectives. Physiol. Plant. 79 : 125-134.

Potter, H. (1988). Electroporation in biology : Methods, Applications and instrumentation. Anal. Biochem, 174 : 361-373.

Raineri, D.M. ; Bottino, P. ; Gordon, M.P. and Nester, E.W. (1990). *Agrobacterium*-mediated transformation of rice (*Oryza sativa* L.). Bio/Technology. 8 : 33-38.

Raleigh, E.A. and Wilson, G. (1986). *Escherichia coli* k-12 restricts DNA containing 5-methylcytosine. Proc. Natl. Acad. Sci. USA. 83 : 9070-9074.

Raleigh, E.A. ; Murray, N.E. ; Revel, H. ; Blumenthal, R.M. ; Westway, D. ; Reith, A.D. ; Rigby, P.W. ; Elhai, J. and Hanahan, D. (1988). McrA and McrB restriction phenotypes of some *E. coli* strains and implications for gene cloning. Nucl. Acid Res. 16 : 1563-1575.

Raleigh, E. (1992). Organization and function of the *mcrBC* genes of *E. coli* K-12. Mol. Microbiol. 6 : 1079-1086.

Rech, E.L. ; Dobson, M.J. ; Davey, M.R. and Mulligan, B.J. (1990). Introduction of a yeast artificial chromosome vector into *Saccharomyces cerevisiae* cells by eletroporation. Nucl. Acids Res. 18 : 1313.

Resis, B. ; Sprengel, R ; Will, H. and Schaller, H. (1984). A new sensitive method for qualitative and quantitative assay for neomycin phosphotransferase in crude cell extracts. Gene. 30 : 211-218.

Revuetta, L. & Jayaram S. (1986). Transformaton of *Phycomyces blakesleeanus to* G-418 resistance by an autonomously replicating plasmid. Proc. Natl. Acad. Sci., USA. 83 : 7344-7347.

Robins, D.M. ; Ripley, S. ; Henderson, A.S. and Axes, R. (1981). Transforming DNA integrates into the host chromosome. Cell 23 : 29-39.

Rogers, S. ; Bisaro, D. ; Horsch, R. ; Fraley, R. ; Hoffmann, N. ; Brand, L. ; Elmer, J. and Lioyd, A. (1986). Tomato golden mosaic virus A component DNA replicates autonomously in transgenic palnts. Cell. 45 : 593-600.

Sandler, S.J. ; Stayton; M. ; Townsend, J.A.; Ralston, M.L. ; Bedbrook, J.R. and Dunsmuir, P. (1988). Plant Mol. Biol. 11 : 301-310.

Sanford, J. ; Klein, T. ; Wolf, E. and Allen, N. (1987). Delivery of substances into cells and tissues using a particle bombardment process. J. Part. Sc. Tech. 5 : 27-37.

Sanford, J. (1988). The biolistic process–a new concept in gene transfer and biological delivery. Trends in Biotechnology. 6 : 299-302.

Sanford, J. (1990a) Biolistic plant transformation–a critical assessment. Physiol. Plant. 79 : 206-209.

Sanford, J. (1990b) The biolistic process–an emerging tool for research and clinical applications. In Milulecky D., Clarke A. (eds.) Proceedings of the biomedical society. Virginia Polytechnic Institute. Blacksburg, VA NY : New York University Press. pp. 89-98.

Sanford, J. ; Smith, F. and Russell, J. (1991). Optimizing the biolistic process for different biological applications. Methods Enzymol. (in Press).

Sawahel, W.A. and Cove, D.J. (1992). Model systems for studying plant molecular biology. Biotechnol. Education J.

Sawahel, W.A. and Cove, D.J. (1992). Gene transfer strategies in plants. Bitechnol. Advances. 10 : 393-412.

Sawahel, W.A. ; Onde, S. ; Knight, C.D. and Cove, D.J. (1992). Transfer of foreign DNA into *Physcomitrella patens* protonemal tissue using the gene gun. Plant Mol. Biol. Rep. 10 : 315-316.

Sawahel, W.A. (1993). Gene transfer strategies. Rice Biotech. Qurt. 14 : 35-37.

Sawahel, W. ; Sastry, G. ; Knight, C. and Cove, D. (1993). Development of an electro-transformation system for *E. coli* DH10B. Biotech. Techniques. 7 : 261-267.

Sawahel, W.A. (1993). Transgenic plants : Performance, Release, and Containment. World J. Of Microbiology and Biotechnology. 10 : 4-11.

Sawahel, W.A. ; Knight, C.D. and Cove, D.J. (1993). Stable genetic transformation of the moss, *Physcomitrella patens* using microprojectile bombardment. J. Experimental Botany.

Sawahel, W. (1994). Transgenic Plant Technology. Rice Bitechnol. Quarterlt. In press.

Sawahel, W.A. ; Knight, C.D. ; Wallace, J. and Cove, D.J. (1994). Co-and re-transformation of the moss *Physcomitrella patens*. J. Experimental Botany. In press.

Soyfer, V. (1980). Hereditary variability of plants under the action of exogenous DNA. Theor. Appl. Genet. 58 : 225-235.

Scangos, G. ; Huttner, K. ; Silver-stein, S. and Ruddle, F.H. (1979). Co-transfer of two linked human genes into cultured mose cells. Proc. Natl. Acad. Sci. USA. 73 : 1274-1278.

Schaffner, W. (1985). In Eukaryotic Transcription : The role of cis- and trans-acting elements in initiation, ed. Gluzman, Y. (Cold Spring Harbor Lab., Cold Spring Harbor, NY), pp. 1-18.

Scheid, O.; Karesch, H. and Potrykus, I. (1990). In activation of a transgene in *Arabidopsis thakana*. In "Abstracts of the Fourth International Congress on Arabidopsis Research' p. 70.

Schiestl, R.H. and Petes, T.D. (1991). Integration of DNA fragments by illegitimate recombination in *Saccharomyces cervisiae*. Proc. Natl. Acad. Sci. USA. 88 : 7585-7589.

Schocher, R. ; Shillito, R.D. ; Saul, M.W. ; Paszkowski, J. and Potrykus, I. (1986). Co-transformation of unlinked foreign genes into plants by direct transfer. Bio/Technology 4 : 1093-1096.

Schwesinger, M.D. (1977). Additive recombination in bacteria. Bacteriol. Ann. Rev. 41 : 872-902.

Shillito, R.D. ; Saul, M.W. ; Paszkowski, J. ; Muller, M. and Potrykus, I. (1985). High efficiency direct gene transfer to plants. Bio/ Technology. 3 : 1099-1103.

Shimamoto, K. ; Terada, R. ; Izawa, T. and Fujimoto, H. (1989). Fertile transgenic rice plants regenerated from transformed protoplasts. Nature. 338 : 274-276

Shirsat, A.; Wilford, N. and Croy, R. (1989). Gene copy number and levels of expression in transgenic plants of a seed specific gene. Plant Sc. 61 : 75–80.

Slightom, J. ; Jouanin, L. ; Leach, F. ; Drong, R. and Tepfer, A. (1985). Isolation and identification of TL-DNA/plant junction in *Convolvulus arvensis* transformed by *Agrabacterium rhizogenous* strain A4. EMBO J., 4 : 3069-3077.

Simoens, C. ; Alliotte, T. ; Mendel, R. ; Muller, A. ; Schiemann, J. ; van Jsebettens, M. ; Schell, J.; van Montagu, M. and Inze, D. (1986). A binary vector for transferring genomic libraries to plants. Nucleic Acids Res. 14 : 8073-8090.

Singer, E. and Beckwith, J. (1966). Transposition of the *lac* region of *E. coli* III. The mechanism of attachment of bacteriophage o 80 ti bacterial chromosome. J. Mol. Biol. 22 : 33-51.

Sodeinde, O. and Kindle, K. (1993). Homologous reombination in the nuclear genome of *Chlamydomonas reinhadrdtii.* Proc. Natl. Acad. Sci. USA. 90 : 9199-9203.

Solomon, A. (1960). Pores in cell membrane. Scientific American, December.

Sompayrac, L.M. and Danna, K.J. (1981). Efficient infection of monkey cells with DNA of simian virus 40. Proc. Natl. Acad. Sci. USA. 78 : 7575-7578.

Spivak, G. ; Ganesan, A.K. and Hanawalt, P.C. (1984). Enhanced transformation of human cells by U.V.–irradiated pSV2. Mol. Cell. Biol. 4 : 1169-1171.

Stacey, D.W. (1981). Microinjection of mRNA and other macromolecules into living cells. Methods in Enzymol. 79 : 76-88.

Steinbiss, H.H. and Stable, P. (1983). Protoplast derived tobacco cells can survive capillary microinjection of the fluorescent dye Lucifer Yellow. Protoplasma. 116 : 223-227.

Stent, G. (1971). Molecular genetics. San Franciso : Freeman.

Stuart, G. ; MacMurry, J. and Westerfield, M. (1988). Replication, integration and stable germ-line transmission of foreign sequences injected into early Zebrafish embryos. Development. 103 : 403-412.

Svab. Z. ; Hajdukiewicz. P. and Maliga, P. (1990). Stable transformation of plastids in higher plants. Proc. Natl. Acad. Sci. USA. 87 : 8526-8530.

Svab. Z. and Maliga, P. (1993). High-frequency plastid transformation in tobacco by selection for a chimeric *aad*A gene. Proc. Natl. Acad. Sci. USA. 90 : 913-917.

Szostak, J.W. ; Orr-weaver, T.L. ; Rothstein, R.T. and Stahi, F.W. (1983). The Double strand break repair model for recombination. Cell. 33 : 25-35.

Tarantul, V.Z. ; Kucheriavy, V.V. ; Makarova, I.V. ; Baranow, Yu. N.; Begetova, T.V. ; Andreeva, L.E. and Gazaryyan, K.G. (1986). Rearrangements of microinjected recombinant DNA in the genome of transgenic mice. Mol. Gen. Genet. 203 : 305-311.

Taylor, W.C. (1989). Regulatory interactions between nuclear and plastid genomes. Annu. Rev. Plant Physiol. Plant Mol. Biol. 40 : 211-233.

Taylor, M.G. and Vasil, I.K. (1991). Histology of, and physical factors affecting, transient GUS expression in pearl millet (*Pannisetum glaucum* (L) R. Br.) embryos following microprojectile bombardment. Plant Cell Rep. 10 : 120-125.

Teeri, T. ; Lehvaslaiho, H. ; Franck, M. ; Uotila, J. ; Heino, P. ; Plava, E. ; van Montagu, M. and Herrera-Estrella, L. (1989). Gene fusions to *LacZ* reveal new expression patterns of chimeric genes in transgenic plants. EMBO J. 8 : 343-350.

Topfer, R. ; Prols, M. ; Schell, J. and Steinbib, H.H. (1988). Transient gene expression in tobacco protoplasts : II. Comparison of the reporter gene systems for CAT, NPTII and GUS. Plant Cell Report. 7 : 255-258.

Topping, J.; Wei, W. and Lindsey, K. (1991) Functional tagging of regulatory elements in the plant genome. Development 112 : 1009–1019.

Tsukakoshi, M. ; Kurata, S. ; Nomiya, Y. ; Ikawa, Y. and Kasuya, T. (1984). A novel method of DNA transfection by laser microbeam cell surgery. Appl. Phys. B. 35 : 135-140.

Twell, D. ; Klein, T. ; Fromm, M. and McCormick, S. (1989). Transient expession of chimeric genes delivered into pollen by microparticle bombardment. Plant Physiol. 91 : 1270-1274.

Uchimiya, H. ; Hirochika, H. ; Hashimoto, H ; Hara, A. ; Masuda, T.; Kasumimoto, T.; Harada, H. ; Ikeda, J. and Yoshioka, M. (1986). Co-expression and inheritance of foreign genes in transformants obtained by direct DNA transformation of tobacco protoplasts. Mol. Gen. Genet. 205 : 1-8.

Van Duin, M. ; Westerveld, A. and Hoeijmarkers, J.H. (1985). U.V. stimulation of DNA-mediated transformation of human cells. Mol. Cell. Biol. 5 : 734-741.

Verhoeven, H.A. and Blass, J. (1988). Direct cell to cell transfer of mitochondria by microinjection. In Proc. of 7th Inter. Protoplast Conf.,Wageningen (K.J. Puite; J.J.M. Dons; H.J. Huizing; A.J. Krol; M. Koornneef & F.A. Kreus, eds.). Kluwer Academic Publishers, Dordrecht. ISBN 90-247-3688-9.

Walden, R. (ed) (1988). Genetic transformation in plants. Open Univ. Press, UK.

Wang, Y. ; Klein, T. ; Fromm, M. ; Cao, J. ; Sanford, J. and Wu, R., (1988). Transformation of rice, wheat and soyabean by the particle bombardment method. Plant Mol. Biol. 11 : 433-439.

Warne, T. and Hickok, L. (1987). Single gene mutants tolerant to NaCl in the fern *Ceratopteris* : Characterization and genetic analysis. Plant Sci. 52 : 49-55.

Weber, G. ; Monajembashi, S. ; Greulich, K. and Wolfrum, J. (1987). Uptake of DNA in chloroplasts of *Brassica napus* (L.) by means of a microfocussed laser beam. Eur. J. Cell Biol. 43 : 63-66.

Weber, G. ; Monajembashi, S. ; Greulich, K. and Wolfrum, J. (1988). Microperforation of plant tissue with a UV-laser microbeam and injection of DNA into cells. Naturwissenschaften. 75 : 35-36.

Weber, G. ; Monajembashi, S. ; Wolfrum, J. and Greulich, K. (1988). A laser micro-beam as a tool to introdue genes into cells and organelles of higher plants. Ber. Bunsenges Phys. Chem. 93 : 252-254.

Weber, G. ; Monajembashi, S. ; Wolfrum, J. and Greulich, K. (1990). Genetic changes induced in higher plant cells by a laser microbeam Physiol. Plant. 79 : 190-193.

Weber, G. and Gruelich, K. (1992). Manipulation of cells, organelles and genomes by laser microbeam and optical trap. Int. Rev. Cytol. 131 : 1-41.

Werr, W. and Lorz, H. (1986). Transient gene expression in a gramineae line. A rapid procedure for studying plant promoters. Mol. Gen. Genet. 202 : 471-475.

Wiegand, R. ; Weber, G. ; Zimmermann, K. ; Monajembashi, S. ; Wolfrum, J. and Geulich, K. (1987). Laser-induced fusion of mammalian cells and plant protoplasts. Journal of cell science. 88 : 145-149.

Wigler, M. ; Sweet, R. ; Sim, G.K. ; Wold, B. ; Pellicer, A. ; Lacy, E. ; Maniatis, T. ; Silverstein, S. and Axel, R. (1979). Transformation of mammalian cells with genes from procaryotes and eukaryotes. Cell. 16 : 777-785.

Willecke, K. and Ruddle, F.H (1975). Transfer of the human gene for Hypoxanthine-Guanine phophoribosy ltransferase via isolated Human metaphsase chromosomes into mouse L-Cells. Proc. Natl. Acad. Sci. USA. 72 : 1792-1796.

Williamson, J.; Hirsch–Wyncott, M.; Larkins, B. and Gelvin, S. (1989). Differential accumulation of a transcript driven by Ca MV 35 S promoter in transgenic tobacco. Plant Physiol. 90 : 1570–1576.

Wilson, J.H. (ed) (1985). Genetic recombination. Annual reviews Cummings Publishing Company Inc.

Wilson. J. ; Leung, W. ; Bosco, G. ; Dieu, D. and Haber, J. (1994). The frequency of gene targeting in yeast depends on the number of target copies. Pro. Natl. Acad. Sci. USA. 91 : 177-181.

Winegar, R.A. ; Phillips, J.W. ; Youngblom, J.H. and Morgan, W.F. (1989). Cell electroporation is a highly efficient method for introducing restriction endonuclease into cell. Mutation Res. 225 : 49-53.

Yamada, Y. and Morikawa, H. (1985). Protoplast fusion of secondary metabolite-producing cells. In Neumann, K. ; Metabolism of plant cell culture, Spring-Verlag, Berlin, pp. 256-267.

Yorifuji, T.S. ; Tsuruta, S. and Mikawa, H. (1989). The effect of cell synchronization on the efficiency of stable gene transfer by electroporation. FEBS lett. 245 : 201-203.

Yorifuji, T.S. and Mikawa, H. (1990). Co-transfer of restriction endonucleases and plasmid DNA into mammalian cells by electroporation : Effect on stable transformation. Mutation Res. 243 : 121-126.

Yu, G.–L. and Blackburn, E.H. (1990). Amplification of tandemly repeated origin control sequences confers a replication advantages on rDNA replicons in *Tetrahymena thermofila.* Mol. Cell. Biol. 10 : 2070-2080.

Zelenin, A. ; Titomirov, A. and Kolesnikov, V. (1989). Genetic transformation of mouse cultured cells with the help of high velocity mechanical DNA injection. FEBS Letters. 244 : 65-67.

Zhang, S. and Dong, W. (1987). Chromosomal aberrations induced by the restriction endonucleases EcoRI, PstI, SalI and BamHI in CHO cells. Mutation Res. 180 : 109-114.

Zhang, L. ; Cheng, L. ; Xu, N. ; Zhao, N. ; Li, C. ; Jing, Y. and Jia, S. (1991). Efficient transformation of tobacco by ultrasonication. Bio/Technology. 9 : 996-997.

Zimmermann, U. ; Pilwat, G. and Riemann, F. (1974). Dielectric breakdown of cell membranes. Biophys. J. 14 : 881-889.

Zimmermann, U. ; Pilwat, G. ; Pequeux, A. & Gilles, R. (1980). Electro-mechanical properties of human erythrocyte membranes : The pressure-dependence of potassium permeability. J. Membrane Biol. 54 : 103-113.